CMS/CAIMS Books in Mathematics

CMS/CAIMS Books in Mathematics is a collection of monographs and graduate-level textbooks published in cooperation jointly with the Canadian Mathematical Society-Societé mathématique du Canada and the Canadian Applied and Industrial Mathematics Society-Societé Canadienne de Mathématiques Appliquées et Industrielles. This series offers authors the joint advantage of publishing with two major mathematical societies and with a leading academic publishing company. The series is edited by Karl Dilcher, Frithjof Lutscher, Nilima Nigam, and Keith Taylor. The series publishes high-impact works across the breadth of mathematics and its applications. Books in this series will appeal to all mathematicians, students and established researchers. The series replaces the CMS Books in Mathematics series that successfully published over 45 volumes in 20 years.

More information about this series at http://www.springer.com/series/16627

CMS
SMC

CAIMS
SCMAI

Andreas Buttenschön • Thomas Hillen

Non-Local Cell Adhesion Models

Symmetries and Bifurcations in 1-D

Andreas Buttenschön
Department of Mathematics
University of British Columbia
Vancouver, BC, Canada

Thomas Hillen
Department of Mathematical
and Statistical Sciences
University of Alberta
Edmonton, AB, Canada

ISSN 2730-650X   ISSN 2730-6518   (electronic)
CMS/CAIMS Books in Mathematics

https://doi.org/10.1007/978-3-030-67111-2

Mathematics Subject Classification: 35R09, 45K05, 35Q92, 92C15, 47G20

This Springer imprint is published by the registered company Springer Nature Switzerland AG
The registered company address is: Gewerbestrasse 11, 6330 Cham, Switzerland

Preface

Whenever cells form tissues, organs, or organisms, they interact with each other through cellular adhesions. Cell–cell adhesions give the skin its stability, they keep cells together to form organs, they allow immune cells to move through the body, and they keep blood inside the vessels. Cell–cell adhesions also facilitate diseases such as cancer and cancer spread through metastasis. A good understanding of this basic cell mechanism is of upmost importance, and here we use mathematics to help to shed light on this interesting process.

We will focus on a mathematical model that was introduced in 2006 by *Armstrong, Painter, and Sherratt*. Their model stands out since it gave the first continuum description of cell adhesion in the form of an integro-partial differential equation (iPDE) model. The model has been used with great success to describe the cell-sorting experiments of Steinberg, as well as tissue dynamics in embryogenesis, wound healing, and cancer cell invasions. Since the iPDE of Armstrong et al. contains non-local terms inside a nonlinear partial differential equation, the mathematical analysis is challenging, even in the "simplest" case of one-cell species in one dimension.

In Part I (*Introduction*) of this monograph, we review applications of the Armstrong, Painter, and Sherratt model to cell aggregations, cell sorting, and cancer spread, and we consider the biological processes that are the basis of this model. We review the known theoretical results on local and global existence and pattern formation. We define an adhesion potential, and we outline a formal connection to *aggregation equations* that have been derived for physical problems, such as the *McKean–Vlasov* model. In fact, many tools from the analysis of the aggregation equations become available for our purpose.

In Part II (*The Periodic Problem*), we develop a full bifurcation theory for the non-local adhesion model in a one-dimensional periodic domain. We combine global bifurcation results pioneered by Rabinowitz, equivariant bifurcation theory, and the symmetries of the non-local term to obtain local and global bifurcation results for the branches of non-trivial solutions. The key idea here is the introduction of a new *area function* in Definition 5.3. Our non-local model does not obey a classical maximum principle, but the area function does, giving us the necessary structure to untangle the bifurcations. We identify general criteria that allow us to distinguish between super- and sub-critical bifurcations.

In Part III (*Non-local Equations with Boundary Conditions*), we extend the non-local cell adhesion model to a bounded domain with no-flux boundary conditions. Since the model is non-local, there is a need to properly define the non-local term once it reaches the boundary. We use biological properties to derive such boundary terms. We propose several different non-local terms incorporating different biologically realities such as boundary adhesion and boundary repulsion. We show that these newly constructed non-local operators are weakly differentiable, using the theory of distributions. In numerical simulations, we see that adhesive boundary conditions lead to boundary *wetting*, quite similar to the behavior of certain fluids. Finally, we use an asymptotic expansion, and numerical simulations to study the steady states of the non-local cell adhesion model incorporating no-flux boundary conditions, and we find a similar bifurcation structure as in the periodic case.

While preparing this work, we were intrigued by the rich mathematical structures of the non-local adhesion model. Not only do these structures allow us to understand its bifurcations, but these structures carry meaning for biological applications. The detailed analysis, as presented here, shows a stimulating interaction between model symmetries, mathematical analysis, and biological reality, which inspired us, and hopefully, our readers as well.

Vancouver, BC, Canada   Andreas Buttenschön

Edmonton, AB, Canada   Thomas Hillen
February 2021

Contents

Part I
Introduction

Chapter 1
Introduction

Cellular adhesion is one of the most important interaction forces in tissues. Cells adhere to each other, to other cells, and to the extracellular matrix (ECM). Cell adhesion is responsible for the formation of tissues, membranes, vasculature, muscle tissue, as well as cell movement and cancer spread [79, 187]. At the molecular level, cellular adhesion is facilitated by a wide range of different cell membrane proteins with integrins and cadherins being the most prominent adhesion molecules [2, 71, 79, 87]. We recognize that cell adhesions are fundamental for the normal functions of organs, embryonic development, wound healing, as well as pathological issues such as cancer metastasis [86, 127, 154].

A good understanding of adhesion and its dynamic properties is essential, and mathematical modelling is one powerful tool to gain such an understanding. There have been several modelling attempts for adhesion, and it turns out that one of the more successful models is the model of Armstrong, Painter, and Sherratt from 2006 [10]. It has the form of a non-local partial differential equation, where the particle flux is an integral term that arises as balance of all the adhesion forces acting on a cell.

The present monograph focuses on mathematical properties of the non-local Armstrong model. The non-local nature of the particle flux term is a challenge, and sophisticated new methods need to be derived. Here we show the existence of non-trivial steady states and analyze their stability and their bifurcation structure. The results are largely based on the abstract bifurcation theory of Crandall and Rabinowitz [156]. We show that the non-local term acts like a non-local derivative, which allows us to define non-local gradients and non-local curvature. Furthermore, we discuss the development of appropriate, and biologically realistic, no-flux boundary conditions, and we show the existence of non-trivial steady states for this case. As the no-flux boundary conditions are non-unique, they open the door for further studies of boundary behavior of cells on tissue boundaries.

© The Author(s), under exclusive license to Springer Nature Switzerland AG 2021
A. Buttenschön, T. Hillen, *Non-Local Cell Adhesion Models*, CMS/CAIMS Books in
Mathematics 1, https://doi.org/10.1007/978-3-030-67111-2_1

1.1 The Effect of Cellular Adhesions in Tissues

Early in the last century, the first biological experimenters had begun to uncover the role of cell adhesion in tissues. One of the earliest observations was that if a sponge is squeezed through a fine mesh [170, 189], it will reform into a functional sponge after transition. A few years later, Hoftreter observed that different tissues have different associative preferences [170]. To describe his observations, he introduced the concept of "tissue affinities". Further, he repeated the earlier observations that previously dissociated tissues have the ability to regain their form and function after deformation (Fig. 1.1). Today, this phenomenon is referred to as cell sorting, and we recognize its critical importance in the formation of functional tissues during organism development.

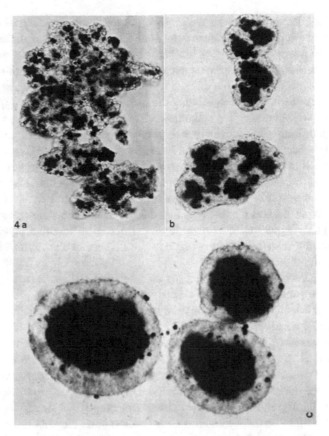

Fig. 1.1 Two cell populations, black and white, with adhesion molecules of different strengths on their surfaces. Initially (**a**) the cells are mixed, and over time they slowly (**b**) re-sort themselves to a sorted final configuration (**c**) (For details on the experimental setup and the figure, see [12])

In 1963, Steinberg proposed the first theory of cell sorting that argued that cell-level properties, namely, a cell's adhesion molecules, drive cell sorting [170, 171]. His theory, capable of explaining the different cell sorting patterns, is known as the *Differential Adhesion Hypothesis* (DAH). Steinberg observed that clusters of cells of the same type behave as if a surface tension holds them together, quite similar to fluid droplets. In other words, cells rearrange to maximize their intracellular attraction and minimize surface tension. That is, the DAH asserts that cell sorting is solely driven by the quantitative differences in the adhesion potential between cell types (e.g., cells with the highest potential of adhesion are found at the center of aggregates). Interesting is that Steinberg referred to adhesion being a "merely close range attraction" [170]. An overview of the experimental verifications can be found in [172].

Further analysis of cell sorting patterns [10, 67] has shown several types of cell clusters as shown in Fig. 1.2. The outcome depends on the relative adhesion strengths of cells of the same type to each other and to cells of a different type.

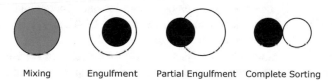

| Mixing | Engulfment | Partial Engulfment | Complete Sorting |

Fig. 1.2 The four possible outcomes of cell sorting with two cell populations. The more cohesive cell population is black. Mixing occurs with preferential cross-adhesion, engulfment with intermediate cross-adhesion, partial engulfment with weak cross-adhesion, and cell sorting with no cross-adhesion [10, 67]

Harris formulated a first critique of the DAH [91]. The main points of his critique were as follows: (1) cells are living objects and thus open thermodynamic systems (not closed as assumed by the DAH); (2) cell size and cell membrane protrusions are much larger than individual adhesion bonds, thus making cellular adhesions a non-local process; and (3) the work of adhesion and de-adhesion may be different, as cells can stabilize adhesion bonds after their formation [91]. To resolve these issues, Harris proposed the *Differential Surface Contraction Hypothesis*, arguing the contractile strength of a membrane completely describes its surface tension. A model similar to this idea was later implemented in a successful vertex model of cell sorting in epithelial tissues [25, 26, 41].

Cellular adhesion is facilitated by a wide range of cell surface molecules and cell-cell junctions [2, 71]. *Adherens junctions* connect the actomyosin networks of different cells with each other, and they form strong bonds and initiate cell polarization; *tight junctions* are the strongest cell-cell connections, and they are used to create impermeable physical barriers; *gap junctions* are intracellular ion channels that allow cell-cell communication; *integrins* connect the cell cytoskeleton with the ECM plus a selection of further adhesion molecules such as *cadherins, igCAMS, Slit/Robo, Ephrin/Eph*, etc., [2, 71, 79, 145, 175].

The transition between tightly packed cells and free-moving cells is fluent, and it depends to a large extent on the adhesion processes that are involved. For example, myoblasts and myofibers are so tightly connected that they can exert physical forces (muscles); epithelial sheets form the lining of many organs and vessels, where it is important to separate an "inside" from an "outside" [79]; in angiogenesis, endothelial cells move and sprout new vessels, which requires loose adhesion as compared to mature vessels [79]; the movement of immune cells occurs in small cell clusters or as individual cells [145], and it requires highly variable adhesion properties of the immune cells. Cancer cells often lose their adhesive properties, which results in local invasion and metastasis [187]. The form of cancer invasion is highly variable, and many different types have been identified, including cluster invasion, small cell groups, cancer-immune cell clusters, individual cells, and network-type invasions [71].

An important process in tissues is the *epithelial-mesenchymal transition* (EMT), where stationary epithelial cells lose their adhesive properties and become invasive, mesenchymal-like cells [86, 139, 187]. The EMT requires a combination of mechanisms such as changed cytoskeletal dynamics and changed adhesive properties. The EMT is a hallmark of cancer metastasis [89, 90], and it is very important to understand the influence of adhesion on the EMT.

A detailed mathematical analysis of adhesion models can contribute to our understanding of this important process and explain, complement, and enrich biological and experimental observations.

1.2 Prior Modelling of Cellular Adhesions

Cell adhesions are forces that act on the cell membrane. For a mathematical or computational model, these forces need to be computed and balanced. Hence an individual cell modelling approach seems to be a natural way to start. Indeed, the first model for cell sorting through differential adhesion was a model of Graner and Glazier from 1991. They used a Cellular Potts model approach, where individual cells are represented as collection of lattice sites in a two-dimensional lattice [80, 83–85]. Cellular adhesions are implemented as interface energies of cells that are touching at a common interface. Since in this model a single cell contains many lattice sites, the adhesion is a non-local interaction. The deformation and movement of the cells is described as an energy minimization approach. Interface energies are balanced with cell volume and cell shape energies plus a random component due to noise. At each step, random changes in the lattice configuration are proposed and accepted by a Boltzmann-like function. Over the years, these Cellular Potts models have become very successful modelling tools, and they have been widely used in applications [165]. For example, Turner et al. [180] used a Cellular Potts model to study the effect of adhesion at the invasion front of a tumor. They observed the formation of clusters of invasion and the formation of "fingering" invasion fronts. In [181], they attempted to scale the Cellular Potts model to a partial differential equation. However, the obtained macroscopic equations are notoriously difficult to analyze.

In 1996, Byrne et al. [32] studied the growth of avascular tumor spheroids in the presence of an external nutrient. The tumor growth is determined by the balance between proliferative pressure and cell-cell adhesion, which keep the spheroid compact. The DAH of Steinberg of surface tension on cell clusters is implemented by the Gibbs-Thompson relation, which relates the tumor spheroid's curvature to the external nutrient concentration. It is assumed that cell-cell adhesion is the force that maintains this curvature [32]. Later, this model was modified such that the cell's proliferation rate depended on the total pressure acting on the cell (due to adhesion and repulsive forces) [33]. This model was then successfully compared to a cell-based model of tumor spheroid growth [33]. In a similar model, Perumpanani et al. [151] introduced a density-dependent diffusion term in a tumor spheroid model; the idea was that cells in high-density areas are slowed down by the presence of adhesion bonds to neighbors. Since then, this approach has been used in more complicated models of tumor growth (see [118, 123]). The adhesive mechanism in these models is purely local. Further, none of these models was able to reproduce cellular aggregations nor cell sorting commonly linked to adhesive interactions.

Different to the Cellular Potts model, Palsson et al. [150] used a lattice-free model, resolving the individual physical forces between the cells using the theory of elasticity. Cells are represented as deformable ellipsoids with long- and short-range interactions with other cells. This model is a non-local individual-based model, and Palsson et al. used it to describe chemotaxis and slug formation in *Dictyostelium discoideum* [150]. In a similar approach, cells are modelled as elastic isotropic spheres, where adhesive and repulsion forces between adhering elastic spheres are resolved using a modified Hertz model [163, 164] or the Johnson, Kendall, and Roberts model [57, 104, 111]. Since these interactions act over a wide range of cell separations, they are non-local models.

Brodland et al. used a vertex model (an individual-based model) to model cell sorting in epithelial tissues [26, 41]. Similar to Steinberg's assumptions, the model considers surface tension at cell-cell interfaces. The surface tension in their model depended on the forces of adhesion, membrane contraction, and circumferential microfilament bundles [26]. They summarized the findings of their numerical studies by formulating the *Differential Interfacial Tension Hypothesis* of cell sorting [25]. Once again, this was a non-local description of cell adhesion.

Since up to this point all cellular adhesion models capable of explaining aggregations and cell sorting were based on non-local models in cell-based approaches, Anderson proposed to combine the continuum and cell-based approaches in a hybrid model [8]. The significance of this hybrid approach is that cells are individually represented (adhesion effects can be taken into account) and environmental factors such as diffusing proteins and chemokines can be modelled using well-established reaction diffusion equations. This approach has been popular in studying the dynamics of tumor spheroids [157, 158].

In 2006, Armstrong et al. proposed the first continuum model of cellular adhesions capable of explaining adhesion-driven cell aggregations and cell sorting [10]. Since this model is the focus of our monograph, we represent it in its basic and one-dimensional form here. Let $u(x, t)$ denote the density of a cell population at spatial location x and time t. Then its evolution subject to random motility and cell-cell adhesion is given by the following non-local integro-partial differential equation

$$\frac{\partial}{\partial t}u(x, t) = \underbrace{D\frac{\partial^2}{\partial x^2}u(x, t)}_{\text{random motility}} - \underbrace{\alpha\frac{\partial}{\partial x}\left(u(x, t)\int_{-R}^{R} h(u(x + r, t))\Omega(r)\, dr\right)}_{\text{cell-cell adhesion}}, \qquad (1.1)$$

where D is the diffusion coefficient, α the strength of the homotypic adhesion, $h(u)$ a possibly nonlinear function describing the nature of the adhesive force, $\Omega(r)$ an odd function, and R the sensing radius of the cell. We give a detailed description of this model and the biological meaning of the terms in Sect. 2.1; see also [31] for a derivation from a stochastic process. An intuitive explanation of the non-local cell-cell adhesion term in Eq. (1.1) is given in Fig. 1.3. The non-local term represents a tug-of-war of the cells on the right and the cells on the left, with the cell at x moving in the direction of largest force. The effect is that cells move up "non-local" gradients of cell population and thus arises the possibility for formation of cell aggregates. The two-population version of Eq. (1.1) was the first continuum model that correctly replicated cell sorting experiments [10]. The significance of these non-local continuum models is that they extend the existing rich toolbox [24, 60, 66, 134, 141] of reaction-diffusion-advection equations to include cell adhesion and thus models of tissues.

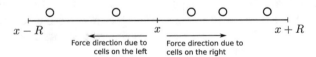

Fig. 1.3 Intuitive description of the non-local adhesion term. Two cells are pulling to the left and three cells are pulling to the right; hence, the net force is to the right (assuming that all cells generate the same force)

As discussed above, cellular adhesions feature prominently in organism development, wound healing, and cancer invasion (metastasis). Therefore, it is unsurprising that model (1.1) has found extensive use in modelling cancer cell invasion [7, 40, 76, 77, 148, 166] and developmental processes [11]. More recently, spatio-temporal variations of the adhesion strengths [56] and adhesion strength variations due to signalling proteins [20] were considered.

The non-local model (1.1) has also been criticized for oversimplification, namely, for its use of a simple diffusion term [133]. Supported by experimental data, Murakawa et al. [133] noticed that under certain conditions Eq. (1.1) gave unrealistic solutions. To address this shortcoming, Murakawa et al. modified the modelling assumption "cells move randomly" to "cells move from high pressure to low pressure regions". For this reason, they introduced a density-dependent diffusion term of porous medium type [133].

Not all adhesion molecules are adhesive; in fact, many such as ephrins are repulsive [149, 175]. These experiments furthermore demonstrate the importance of these repulsive effects in tissue segregation and the creation of sharp boundaries between adjacent tissues. The non-local model (1.1) is easily extended by considering both attractive and repulsive potentials within the non-local integral term. Such extensions have been studied in [15, 16, 146]. Finally, the existence of ephrins on domain boundaries [149] motivates us to develop no-flux boundary conditions for Eq. (1.1) in Chapter 6.

Prerequisite for the extensive numerical exploration of the solutions of the non-local equation (1.1) was the development of efficient numerical methods to evaluate its integral term. An efficient method based on the fast Fourier transform was developed in [75]. Using this efficient algorithm, numerical solutions of Eq. (1.1) are implemented using a spatial finite-volume discretization and the method of lines for the temporal advancement [74]. More recently structure-preserving finite-volume schemes are emerging for (i)PDEs [13, 14, 36]. These schemes guarantee the preservation of the gradient flow structure many scalar iPDEs possess. This allows us to confidently use these schemes in studying the asymptotic dynamics and nonlinear stability of stationary solutions of (i)PDEs.

Existence results for the solutions of the non-local equation (1.1) and generalizations in any space dimension were developed in [7, 20, 102, 166]. Most significant is the general work [102], who showed local and global existence of classical solutions (see also Sect. 2.5). For the case of small adhesion strength, travelling wave solutions of Eq. (1.1) were found and studied in [144].

1.3 Non-local Partial Differential Equation Models

Having introduced the non-local adhesion model (1.1), we take a look at developments of non-local partial differential equation models in general. Non-local models basically arise in two different situations: Firstly, one assumes a priori the existence of non-local interactions, or, secondly, a non-local term arises after solving a partial differential equation or taking some asymptotic limit of it [110, 114, 129]. For example, in a system of four partial differential equations (two cell populations and two diffusing signalling molecules), Knutsdottir et al. [114] applied a quasi-steady-state assumption to the signalling molecules, which diffuse much faster than the cellular populations. The resulting elliptic equations were solved using Green's functions, and thus non-local terms were introduced in the remaining cell density equations. Similar methods are standard in the analysis of chemotaxis (see Horstmann [105]).

Non-local models arise in the study of Levi walks. Levi walks have been used to describe anomalous diffusion, for example, in situations where particles show distinct searching and foraging behavior. A continuum description of these random walks leads to fractional Laplacian operators, which are non-local operators [113, 176].

In discrete time, integro-difference equations are a common tool in ecological modelling, in particular in situations where distinct generations can be identified, and only certain parts of the population spread spatially [116, 138]. One example is dispersal of seeds released annually during a certain season.

Nonlinear integro-partial differential equations have also been derived for birth-jump processes [99]. These are continuous in time and space, and newborn particles are allowed to spread non-locally. Birth-jump processes have been applied to cancer spread [22, 52, 98], to forest fire spread [124], and to the modelling of evolution and selection [42].

The non-local cell-cell adhesion model (1.1) has the non-local term in the advection term. The first non-local equations with the non-local term contained in the advection term were proposed in a series of papers by Nagai et al. in 1983 [135–137]. Their introduction of the non-local term was driven by a desire to model aggregation processes in ecological systems. For comparison with the non-local adhesion model (1.1), the equations of Nagai et al. looked like this

$$u_t = (u^m)_{xx} - \left[u \left\{ \int_{-\infty}^{\infty} K(x-y)u(y,t)\,\mathrm{d}y \right\} \right]_x, \quad x \in \mathbb{R},\ t > 0, \qquad (1.2)$$

where $m > 1$. Shortly after, Alt studied generalizations of Eq. (1.2) in [4, 5]. A version of Eq. (1.2) with finite integration limits and with the special choice of $K(x-y) = \mathrm{sgn}(x-y)$ was studied by Ikeda in [108, 109]. Ikeda established the existence of solutions of Eq. (1.2) on an unbounded domain and developed spectral results in the special case of $m = 2$ that was used to give a classification of the steady states of Eq. (1.2). In 1999, Mogilner et al. [130] used such non-local models to develop evolution equations describing swarms. More recently, such models were used to describe the aggregation of plankton [1] and to model animal populations featuring long-ranged social attractions and short-ranged dispersal [177]. Eftimie et al. [61, 62] used a Lagrangian formulation to obtain non-local hyperbolic models of communicating individuals. Since then, equivariant bifurcation theory was used to study the possible steady states of such communication models [28]. Very recently they discussed the use of Lyapunov-Schmidt and center-manifold reductions study the long-time dynamics of such equations. The non-local adhesion model falls also in this category of equation [10], and more recently it was generalized to include both aggregations and repulsive behavior [146].

Another famous non-local model for biological applications is the *aggregation equation*

$$u_t = \nabla \cdot (u\nabla(u + W * u)), \qquad (1.3)$$

where $W(x)$ is a species interaction kernel that acts non-locally through convolution $*$. The interaction kernel W can be used to describe species aggregation, repulsion, and alignment [177]. The model has been successfully employed to describe animal swarming as well as non-biological applications such as granular matter, astrophysics, semiconductors, and opinion formation (see references in [29]). The aggregation equation (1.3) can be seen as gradient flow of an appropriate energy functional, which contains the non-local interaction potential W. This makes the vast resource of variational methods available to the analysis of (1.3). Many results are known about aggregations, pattern formation, finite time blow-up, and the relation of (1.3) to chemotaxis models [18, 29, 58]. As we show in Sect. 2.6, the adhesion model (1.1) studied here can be formulated as an aggregation equation with linear diffusion term, also known as Mckean-Vlasov equation [37]. The rich methodology of variational calculus becomes available, which we discuss in some detail later.

A non-local model for chemotaxis was proposed in [100, 143] to replace the chemical gradient in a chemotaxis model by the following non-local gradient:

$$\overset{\circ}{\nabla}_R v(x) := \frac{n}{\omega_D(x)R} \int_{\mathbb{S}_D^{n-1}} \sigma v(x + R\sigma) \, d\sigma, \tag{1.4}$$

$\mathbb{S}_D^{n-1}(x) = \{\sigma \in \mathbb{S}^{n-1} : x + \sigma R \in D\}$, and $\omega_D(x) = |\mathbb{S}_D^{n-1}(x)|$. It was shown in [100] that this non-local gradient leads to globally bounded solutions of the non-local chemotaxis equation in cases where the local gradient gives rise to blow-up solutions. The steady states of the local chemotaxis equation [161, 184], chemotaxis with volume filling [35], and the non-local chemotaxis equation [192] have been studied using bifurcation techniques in one dimension.

Non-local terms have also been considered in reaction terms, i.e., in equations of the form

$$u_t = u_{xx} + f(x, u, \bar{u}),$$

where

$$\bar{u} = \int g(x, u) \, dx.$$

These types of equations were studied in [23, 49–51, 68–70]. Oftentimes such systems occur as the asymptotic limit ("shadow system") of a system of reaction-diffusion equations. An application of such a model is, for example, to Ohmic heating in [119]. A different application considers the growth of phytoplankton in the presence of light and nutrients [193]. Yet another application of such a model is a study on the effect of crop raiding of large-bodied mammals [106].

Many of the abovementioned examples are formulated on infinite domains, to avoid subtleties in dealing with boundary conditions or subtleties in ensuring the non-local term is well-defined. Indeed, even the analysis of local equations such as the viscous Burger equation on bounded domains has remained unaddressed until recently [185].

1.4 Outline of the Main Results

In this monograph, we consider non-local models of cell-cell adhesion in the form of an integro-partial differential equations (see Eq. (1.1)). The central problems which we would like to address in this monograph are:

1. What are the non-trivial steady states of the non-local cell adhesion model (i.e., Eq. (1.1)) in a periodic domain?
2. What are their bifurcation structure and their stabilities?

3. How to include biologically realistic boundary conditions that can describe no-flux boundaries, boundary adhesion, or boundary repulsion?
4. What are the non-trivial steady states if boundary conditions are included?

In Chap. 2, we give a brief summary of the derivation of non-local cell-cell adhesion models from an underlying stochastic random walk and summarize the key bifurcation results we employ.

In Chap. 3, we define the non-local term in (1.1) as integral operator $\mathcal{K}[u]$ and explore its mathematical properties. In particular, we establish its continuity properties, L^p-estimates, compactness, and spectral results. These results are key to the subsequent bifurcation analysis. In fact we show that $\mathcal{K}[u]$ is a generalization of the classical first-order derivative. Finally, we establish elementary results on the steady states of the non-local adhesion model (1.1), including an a priori estimate.

In Chaps. 4–5, we investigate the steady states of a single population non-local model of cellular adhesions on a periodic domain. We combine global bifurcation techniques pioneered by Rabinowitz, equivariant bifurcation theory (the equation is $\mathbf{O}(2)$-equivariant), and the mathematical properties of the non-local adhesion term, to obtain the existence of unbounded global bifurcation branches of non-trivial solutions. In words, our main theorem is:

Theorem 1.1 *For each $\alpha > \alpha_n$ (α_n eigenvalues of the **linearized** problem), the periodic non-local linear adhesion steady-state problem (1.1) (i.e., $h(u) = u$) has at least two n-spiked solutions (one in each of $\mathcal{S}_n^{\pm}$).*

The solution branches are classified by the derivative's number of zeros (i.e., the number of extrema remains fixed along a bifurcation branch) or in other words by their number of spikes. The significance of this result is that it parallels the seminal classification of solutions of nonlinear Sturm–Liouville problems [45] and the classification for equivariant nonlinear elliptic equations [94].

In Chap. 6, we consider the construction of no-flux boundary conditions for the non-local cell adhesion model, and finally we explore what we can say about its steady states. In the past, boundary conditions for non-local equations were avoided, because their construction is subtle and requires biological insight. Using the insights from [31], we construct a non-local operator, which takes various boundary effects into account, such as no-flux, or boundary adhesion, or boundary repulsion. These boundary conditions destroy some of the symmetry properties that were available in the periodic case, and consequently, the Rabinowitz-bifurcation theory does no longer apply. We find similar steady states as in the periodic case, but the full bifurcation structure in this case requires new methods. Numerically, we find that steady states show boundary behavior, which is well known from fluids that are wetting the boundary (adhesive) or are repulsed from the boundary.

Chapter 2
Preliminaries

In this section, we present some basic results that are needed later. We give a summary of the derivation of the non-local adhesion model from biological principles as presented in [31], we introduce some notations and methods from abstract bifurcation theory as it was developed in [121, 122], we introduce an averaging operator on periodic domains, and we cite a global existence result for (1.1) in $\mathbb{R}^n$ from [102]. Moreover, in Sect. 2.6, we define an adhesion potential for our model, making an explicit connection to the aggregation equation (1.3).

2.1 Biological Derivation of the Non-local Adhesion Model

The goal of this section is to summarize the derivation of the non-local adhesion model given in Eq. (1.1) from a stochastic position jump process. The derivation of population-level models from underlying mesoscopic movement models has a rich history [34, 101, 141, 142, 173].

In order to focus on the modelling of the cell-cell interactions in the absence of boundary effects, we carry out the derivation on an unbounded domain, i.e., $\Omega = \mathbb{R}^n$. Note that in Chap. 6, we discuss extensions to include boundary effects. Key to the derivation is that cells send out many membrane protrusions to sample their environment. We assume that these membrane protrusions are more frequent than translocations of the cell body; the cell body includes the cell nucleus and most of the cell's mass. In other words, we are most interested in translocations of the cell body and not the frequent, but temporary, shifts due to membrane protrusions. For these reasons, we define the population density function $u(x, t)$ as follows:

$$u(x, t) \equiv \text{Density of cells with their cell body centered at } x \text{ at time } t$$

A. Buttenschön, T. Hillen, *Non-Local Cell Adhesion Models*, CMS/CAIMS Books in Mathematics 1, https://doi.org/10.1007/978-3-030-67111-2_2

We make two assumptions about the movement of the cells:

Modelling Assumption 1 We assume that in the absence of spatial or temporal heterogeneity, the movement of individual cells can be described by Brownian motion. It has been shown that this is a reasonable assumption for many cell types [19, 131, 162].

Modelling Assumption 2 The cells' polarization may be influenced by spatial or temporal heterogeneity. We denote the polarization vector by $\mathbf{p}(x)$.

We describe the evolution of the cell density $u(x, t)$ using a position-jump process, using stochastically independent jumpers (i.e., a continuous-time random walk, for which we assume the independence of the waiting time distribution and spatial redistribution). Here, the waiting time distribution is taken to be the exponential distribution, with constant mean waiting time. Let $T(x, y)$ denote the rate for a jump from $y \to x$ with $x, y \in \Omega$. The evolution of $u(x, t)$ is given by the continuous-time master equation.

$$\frac{\partial u}{\partial t}(x, t) = \lambda \int_D [T(x, y)u(y, t) - T(y, x)u(x, t)] \, d\mu(y), \qquad (2.1)$$

where (D, μ) is a measure space of a physical space (e.g., a domain or a lattice), λ jump rate, and $T(x, y)$ probability density for jump from y to x. For more details on the derivation of the master equation, see [107, 142, 182].

For notational convenience, we associate to a jump from $y \in \Omega$ to $x \in \Omega$ the *heading* $z := x - y$. Using the heading, we define $T_y(z) := T(y + z, y) = T(x, y)$. Let D^y denote the set of all possible headings from y. We assume that the set D^y is symmetric (i.e., if $z \in D^y$, then so is $-z \in D^y$). We further assume that for every $y \in \Omega$, the function T_y is non-negative as it represents a rate.

Given $y \in \Omega$, we denote the redistribution kernel at this location by $T_y(z)$; we assume that $T_y \in L^1(D^y)$ and that $\|T_y\|_1 = 1$ holds. This turns T_y into a probability density function (pdf) on D^y.

Any function which is defined for both z and $-z$ can be decomposed into even and odd components, which are denoted by S_y and A_y, respectively.

Lemma 2.1 (Lemma 1 in [31]) *Consider $y \in \Omega$; given $T_y \in L^1(D^y)$, then there exists a decomposition as*

$$T_y(z) = \begin{cases} S_y(z) + A_y(z) \cdot \frac{z}{|z|} & \text{if } z \neq 0 \\ S_y(z) & \text{if } z = 0 \end{cases} \qquad (2.2)$$

with $S_y \in L^1(D^y)$ and $A_y \in \left(L^1(D^y)\right)^n$. The even and odd parts are symmetric such that

$$S_y(z) = S_y(-z) \quad \text{and} \quad A_y(z) = A_y(-z). \qquad (2.3)$$

Using this decomposition, we define two properties which are analogous to Modelling Assumptions 1 and 2 above. First, we define the motility.

Definition 2.1 (Motility) We define the motility at $y \in \Omega$ as

$$M(y) := \int_{D^y \setminus \{0\}} T_y(z) \, dz = \int_{D^y \setminus \{0\}} S_y(z) \, dz, \tag{2.4}$$

where the integration is w.r.t. the measure on D^y.

The motility is the probability of leaving y. This probability is 1 if $0 \notin D^y$; it is also 1 if $0 \in D^y$ and T_y is a continuous pdf, and it may be smaller than 1 if $0 \in D^y$ and T_y is a discrete pdf. Here we find that the motility depends solely on the even component S_y, in other words solely on Modelling Assumption 1.

Secondly, we define the polarization vector in a *space-jump* process.

Definition 2.2 (Polarization Vector) The polarization vector at $y \in \Omega$ is defined as

$$E(y) := \frac{\int_{D^y} z \, T_y(z) \, dz}{\| \int_{D^y} z \, T_y(z) \, dz \|} = \frac{\int_{D^y} z \, A_y(z) \cdot \frac{z}{|z|} \, dz}{\| \int_{D^y} z \, A_y(z) \cdot \frac{z}{|z|} \, dz \|}. \tag{2.5}$$

where the integration is w.r.t. measure on D^y.

The first moment of the pdf T_y can be intuitively understood as the expected heading of a jump originating at y. This is in direct correspondence with a polarized cell which, following polarization, moves in the direction of the polarization vector. The expected heading is solely determined by A_y, which therefore plays the role of the polarization vector $\mathbf{p}(y)$ in a space-jump process. This correspondence motivates us to set $A_y = \mathbf{p}(y)$ in the subsequent derivations.

We can now derive the macroscopic limit of the master equation (2.1). In addition, to the prior assumptions, we assume that we have a *myopic* random walk such that the jump probability only depends on the jump location y but not on the target location $y + z$. We only consider small jumps of length $h \ll 1$, and we expand (2.1) in h. For the mathematical details, we refer the reader to [31]. At the end of this process, we obtain

$$u_t(x, t) \approx \frac{\lambda h^{n+1}}{2+n} |\mathbb{S}^{n-1}| \Delta(S_x u(x, t)) - \frac{\lambda h^n}{n} |\mathbb{S}^{n-1}| \nabla \cdot (A_x u(x, t)). \tag{2.6}$$

We then assume that $\mathbf{A}_x \sim O(h)$ and let

$$D(x) = \lim_{\substack{h \to 0 \\ \lambda \to \infty}} \lambda h^2 S_x, \qquad \alpha(x) = \lim_{\substack{h \to 0 \\ \lambda \to \infty}} 2\lambda h^2 \mathbf{A}_x.$$

The advection-diffusion limit is

$$\frac{\partial u}{\partial t}(x, t) + \nabla \cdot (\alpha(x) u(x, t)) = \Delta(D(x) u(x, t)). \tag{2.7}$$

We notice that the diffusive part is defined through the symmetric component S_x of T_x, while the advective part α is defined through the anti-symmetric component

$\mathbf{A}_x$. We identify the anti-symmetric part with the polarization vector $\mathbf{A}_x \sim \mathbf{p}$, and it remains to find a good model for the polarization vector $\mathbf{p}$. Here we assume that the cell's polarization vector is determined by the interactions of adhesion molecules on cell's protrusions and adhesion molecules present in the surrounding environment (e.g., located on the surfaces of other cells).

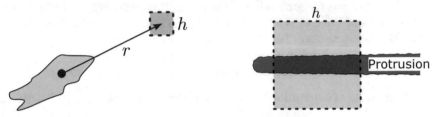

Fig. 2.1 Left: a cell of spatial extend and a small test volume V_h at a distance r from the cell body. Right: a cell protrusion overlapping a small test volume V_h

To describe the microscopic interactions between a cell's protrusion and its surroundings, we consider a small test volume V_h (see Fig. 2.1). We assume that the factors determining the size of contribution of the interactions in V_h to the cell's overall polarization are:

1. The distance from V_h to the cell body is $|r|$.
2. The direction of generated force is $r/|r|$.
3. The free space $f(r)$.
4. The part of protrusion in V_h is called $\omega(r)$.
5. The density of formed adhesion bonds is N_b.
6. The adhesive strength per bond is γ.

The component of the cell polarization that is generated through forces in the test volume V_h is then

$$\mathbf{p}_h(x+r) = \gamma \underbrace{h^n N_b(x+r)}_{\text{\# of adhesion bonds}} \underbrace{f(x+r)}_{\text{free space}} \underbrace{h\omega(r)}_{\text{amt. of cell in } V_h} \underbrace{\frac{r}{|r|}}_{\text{direction}}$$

Summing over all test volumina, we obtain the cell's polarization:

$$\mathbf{p}_{\text{net}}(x) = \int_{\mathbb{B}_R(0)} N_b(x+r)f(x+r)\mathbf{\Omega}(r)\,\mathrm{d}r, \tag{2.8}$$

where $\mathbf{\Omega}(r) = \omega(r)\frac{r}{|r|}$. Finally we let $\alpha(x) = \mathbf{p}_{\text{net}}(x)$, and Eq. (2.7) becomes

$$\frac{\partial u(x,t)}{\partial t} = \nabla \cdot [\nabla (D(x)u(x,t)) - \alpha(x)u(x,t)\mathbf{p}_{\text{net}}(x)], \tag{2.9}$$

with the non-local term $\mathbf{p}_{\text{net}}$ from (2.8).

We are now left with choosing the functions $f(\cdot)$ and the reaction kinetic yielding the density of bound adhesion molecules $N_b(\cdot)$.

The formulation of $\mathbf{p}_{net}$ as in (2.8) allows for the inclusion of detailed biological modelling. We illustrate this in a few examples in one dimension.

Example 1: Armstrong model. One particular choice leading to the original $1-D$ model by Armstrong et al. is:

1. Let $\omega(r)$ be the uniform distribution, i.e., $\Omega(r) = \mathrm{sgn}(r)/2R$.
2. Assume there is always free space $f(x) \equiv 1$.
3. Assume that the background adhesion bond density is proportional to the population density: $N_b(x) \propto u(x)$.

With those assumptions, Eq. (2.9) becomes

$$\frac{\partial u(x,t)}{\partial t} = \frac{\partial}{\partial x}\left[\frac{\partial}{\partial x}(D(x)u(x,t)) - \alpha u(x,t) \int_{-R}^{R} u(x+r,t)\,\Omega(r)\,dr \right].$$

This case is illustrated on the left in Fig. 2.2.

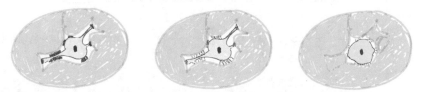

Fig. 2.2 Left: illustration of the Armstrong adhesion model. The cell extends protrusions to make contact with a gray background cell population. The background population is not separated into individual cells and appears only as a density field $u(x,t)$. Middle: in the Bell surface kinetics model, the binding and unbinding of cell surface receptors are modelled explicitly. The cell surface receptors are indicated as blue dashes. Right: the volume filling model with no free space to extend cell protrusions

Example 2: Bell kinetics. If we like to model the binding and unbinding of cell surface adhesion molecules in more detail, we can use the Bell kinetics model [17]. In extension of the previous case, we assume now that the number of adhesion bonds is given by

$$N_b(x) \propto \frac{Ku(x)}{1 + Ku(x)},$$

where K denotes the ratio of association and dissociation constants.

Then Eq. (2.9) in one dimension becomes

$$\frac{\partial u(x,t)}{\partial t} = \frac{\partial}{\partial x}\left[\frac{\partial}{\partial x}(D(x)u(x,t)) - \alpha u(x,t) \int_{-R}^{R} \frac{Ku(x+r,t)}{1 + Ku(x+r,t)}\,\Omega(r)\,dr \right].$$

This case is illustrated in the middle figure of Fig. 2.2. Of course, other binding dynamics can be included in a similar way.

Example 3: Volume filling I. In tightly packed tissue, the available space is limited. This can be included in various ways. Here we assume that the ability to extend protrusions decreases as the tissue becomes packed. This case is illustrated in the right in Fig. 2.2. Hence, compared to Example 1 above, we change the assumption for free space as

$$f(x) = \max\left\{1, 1 - \frac{u(x)}{U_{\max}}\right\}.$$

With those assumptions, Eq. (2.9) in one dimension becomes

$$\frac{\partial u(x,t)}{\partial t} = \frac{\partial}{\partial x}\left[\frac{\partial}{\partial x}\left(D(x)u(x,t)\right) - \alpha u(x,t)\int_{-R}^{R} u(x+r,t)\left(1 - \frac{u(x+r,t)}{U_{\max}}\right)^{+}\Omega(r)\,dr\right].$$

Example 4: Volume filling II. The left figure in Fig. 2.3 indicates a situation where the cell is still able to extend protrusions, but even if they bind, the cell can still not move, as the openings are too small. In one dimension, this form of volume filling is modelled as a multiplicative term in the adhesive flux as

$$\frac{\partial u(x,t)}{\partial t} = \frac{\partial}{\partial x}\left[\frac{\partial}{\partial x}\left(D(x)u(x,t)\right) - \alpha\left(1 - \frac{u(x+r,t)}{U_{\max}}\right)^{+} u(x,t)\int_{-R}^{R} u(x+r,t)\,\Omega(r)\,dr\right].$$

Fig. 2.3 Left: the volume filling case where cells can still make contact with cells that are further away, but there is no space for the cell to move. Right: in this case, the neighboring cells are depicted individually, each having their own size and their own filopodia

Example 5: Non-local background population. So far the background population was treated as a background field of cells. If we want to include more detail of the spatial extent of the background cells, as shown on the right in Fig. 2.3, we can also include their non-local nature by using

$$N_b(x) \propto \int_{-R}^{R} u(x+r)\eta(r)\,dr.$$

Then we obtain a doubly non-local model

$$\frac{\partial u(x,t)}{\partial t} = \frac{\partial}{\partial x} \left[\frac{\partial}{\partial x} \left(D(x)u(x,t) \right) - \alpha u(x,t) \int_{-R}^{R} \int_{-R}^{R} u(x+y+r,t) \, \eta(y) \, dy \, \Omega(r) \, dr \right].$$

This case of a doubly non-local term has, as far as we know, never been studied in detail.

As these examples show, our general formulation allows the inclusion of many detailed biological features, such as receptor binding kinetics, competition for space, spatial extent of background cells, etc. (see [31]). Also, extensions to higher space dimensions and to several cell types are straightforward (see, e.g., [7, 10, 102, 166]).

2.2 Introduction to Nonlinear Analysis

This section is based on the abstract bifurcation theory as developed by J. López-Gómez in [121, 122] and is meant as a quick introduction to the abstract framework which will be employed in this work.

Banach spaces and their subspaces will be denoted using capital letters, that is, X, Y, U, V, and so on. Operators between function spaces will be denoted using the calligraphic font, for example, $\mathcal{L}, \mathcal{F}$, and $\mathcal{K}$. The argument of an operator will be enclosed in square brackets. For instance,

$$\mathcal{L} : X \mapsto Y, \quad \mathcal{L}[x] = y.$$

This is to distinguish the action of the operator from a family of operators. For example, a family of operators may be indexed using a real number λ. Then, we have a map from $\mathbb{R} \to \mathfrak{L}(X, Y)$

$$\lambda \mapsto \mathcal{L}(\lambda) : X \mapsto Y,$$

and for each fixed λ, we may study $\mathcal{L}(\lambda)[x] = y$. The kernel and range of an operator are denoted $N[\mathcal{L}]$ and $R[\mathcal{L}]$.

Spaces of operators are denoted using the Fraktur font. The most important is the space of continuous linear operators denoted $\mathfrak{L}$ and the space of compact operators $\mathfrak{K}$. The space of Fredholm operators is denoted Fred_i where i denotes the index.

Special subspaces such as a continuum of solutions or the solution set of an operator equation are also denoted using the Fraktur font, for example, $\mathfrak{C}$ and $\mathfrak{S}$.

The following sections are based on [121, 122]. Let U and V be two real Banach spaces. We denote the space of bounded linear operators from U to V by $\mathfrak{L}(U, V)$ and by $\text{Fred}_0(U, V)$ the subset of $\mathfrak{L}(U, V)$ containing all Fredholm operators with index 0. The set of all isomorphisms between U and V is denoted $\text{Iso}(U, V)$. The operator $\mathcal{L}$ is said to be Fredholm whenever

$$\dim N[\mathcal{L}] < \infty, \quad \text{codim} \, R[\mathcal{L}] < \infty,$$

and recall that

$$\operatorname{codim} R[\mathcal{L}] = \dim V/R[\mathcal{L}] = \dim \operatorname{coker}[\mathcal{L}].$$

The index of a Fredholm operator is defined by

$$\operatorname{ind}[\mathcal{L}] = \dim N[\mathcal{L}] - \operatorname{codim} R[\mathcal{L}].$$

The most important example of a Fredholm operator with index zero is the following.

Theorem 2.1 (Theorem 4.2 [81]) *Let $\mathcal{K} \in \mathfrak{K}(U)$ (compact operators); then $\mathcal{F} = I - \mathcal{K}$ is a Fredholm operator of index zero.*

Theorem 2.2 (Theorem 4.1 [81]) *Let $\mathcal{L} \in \mathfrak{L}(U, V)$ be a Fredholm operator, and let $\mathcal{K} \in \mathfrak{K}(U, V)$ (compact). Then $\mathcal{F} = \mathcal{L} + \mathcal{K}$ is a Fredholm operator with*

$$\operatorname{ind}(\mathcal{L} + \mathcal{K}) = \operatorname{ind}(\mathcal{L}).$$

When a dynamical system depends on parameters, then it is natural to talk about operator families. In [121, 122], this approach is used to define bifurcation values as values in a generalized spectrum of an operator family as defined below. We feel that the notion of a generalized spectrum is confusing, and we noticed that readers tend to confuse them for the spectrum of an operator. To make this distinction a bit more apparent, we change the original definition and use *singular value, singular point* and *bifurcation point*.

Definition 2.3 Let U and V be two Banach spaces over the field $\mathbb{K}$ and $r \in \mathbb{N}$.

1. An *operator family* $\{\mathcal{L}(\alpha)\}_{\alpha \in \Gamma}$ of class C^r with $\Gamma \subset \mathbb{K}$ from U to V is a map

$$\mathcal{L} \in C^r(\Gamma, \mathfrak{L}(U, V)).$$

   In our application, we have $\alpha \in \Gamma \subset \mathbb{R}$, where α will become our bifurcation parameter.
2. Let $\mathcal{L} \in C(\Gamma, \mathfrak{L}(U, V))$ be an operator family; then the point $\alpha_0 \in \Gamma$ is a *singular value* of $\mathcal{L}$ if

$$\mathcal{L}_0 := \mathcal{L}(\alpha_0) \notin \operatorname{Iso}(U, V),$$

   and it is a *singular point* of $\mathcal{L}$ if

$$\dim N[\mathcal{L}_0] \geq 1.$$

   A singular point α_0 is simple whenever

$$\dim N[\mathcal{L}_0] = 1.$$

3. The set of all singular values of the operator family $\mathcal{L}$ is denoted by

$$\operatorname{Sing} = \operatorname{Sing}(\mathcal{L}) = \{\, \alpha \in \Gamma : \mathcal{L}(\alpha) \notin \operatorname{Iso}(U, V) \,\}.$$

Similarly, the set of all singular points of the operator family $\mathcal{L}$ is defined by

$$\text{SingPt} = \{\, \alpha \in \text{Sing} : \dim N[\mathcal{L}(\Gamma)] \geq 1 \,\}.$$

4. The *structurally stable set* (sss) of $\mathcal{L}$ is defined by

$$\text{sss}(\mathcal{L}) := \Gamma \setminus \text{Sing}.$$

Remark 2.1 Since $\mathcal{L} \in C(\mathbb{R}, \mathfrak{L}(U,V))$ and $\text{Iso}(U,V)$ is an open subset of $\mathfrak{L}(U,V)$, we have that $\text{sss}(\mathcal{L})$ is open and possibly empty. Thus $\text{Sing}(\mathcal{L})$ is closed.

Lemma 2.2 *If $\mathcal{L}_0 \in \text{Fred}_0(U,V)$, then $\alpha_0 \in \text{Sing}(\mathcal{L})$ if and only if $\alpha_0 \in \text{SingPt}(\mathcal{L})$.*

Proof We only have to prove one direction. Let $\alpha_0 \in \text{Sing}(\mathcal{L})$; then $\mathcal{L}_0 : U/N[\mathcal{L}_0] \mapsto R[\mathcal{L}_0]$ is an isomorphism by the open mapping theorem. As $\mathcal{L}_0$ is Fredholm with index zero, we have that $\dim N[\mathcal{L}_0] < \infty$; hence, $\alpha_0 \in \text{SingPt}(\mathcal{L})$. $\square$

Hence, if $\mathcal{L}(\Gamma) \subset \text{Fred}_0(U,V)$, then $\text{Sing}(\mathcal{L}) = \text{SingPt}(\mathcal{L})$.

Remark 2.2 Note that the concept of a singular value and singular point is quite closely related to the standard notations of the spectrum and point spectrum of an operator. In fact, this is the basis for the notion of *generalized eigenvalues*, which we will not use here. However, given an operator A, we can always define an operator family as

$$\mathcal{L}(\lambda) = A - \lambda I$$

Then

$$\text{Sing}(\mathcal{L}) = \sigma(A), \qquad \text{and} \qquad \text{SingPt}(\mathcal{L}) = \sigma_p(A),$$

where $\sigma_p(A)$ denotes the point spectrum of A. In the same way, we could generalize the continuous spectrum and the residual spectrum for operator families. But those will be of no further use in our bifurcation analysis.

2.3 Abstract Bifurcation Theory

This section gives an overview of the central bifurcation results we will apply in this book. For more complete introductions, we refer the reader to [27, 120, 169].

Let U and V be two real Banach spaces, and suppose that we want to analyze the structure of the solution set of the nonlinear operator given by

$$\mathcal{F}[\alpha, u] = 0, \quad (\alpha, u) \in \mathbb{R} \times U,$$

where

$$\mathcal{F} : \mathbb{R} \times U \mapsto V,$$

is a continuous map satisfying the following requirements:

(F1) For each $\alpha \in \mathbb{R}$, the map $\mathcal{F}[\alpha, \cdot]$ is of class $C^1(U, V)$ and

$$\mathcal{D}_u \mathcal{F}(\alpha, u) \in \text{Fred}_0(U, V) \quad \text{for all } u \in U.$$

(F2) $\mathcal{D}_u \mathcal{F} : \mathbb{R} \times U \mapsto \mathfrak{L}(U, V)$ is continuous.
(F3) $\mathcal{F}[\alpha, 0] = 0$ for all $\alpha \in \mathbb{R}$.

Definition 2.4 Assume $\mathcal{F}$ satisfies (F1)–(F3).

1. A *component* $\mathfrak{C}$ is a closed and connected subset of the set

$$\mathfrak{S} = \{ (\alpha, u) \in \mathbb{R} \times U : \mathcal{F}[\alpha, u] = 0 \},$$

  that is maximal with respect to inclusion.
2. As $(\alpha, 0)$ is a known zero, it is referred to as the *trivial state*.
3. Given $\alpha_0 \in \mathbb{R}$, it is said that $(\alpha_0, 0)$ is a *bifurcation point* of $\mathcal{F} = 0$ if there exists a sequence $(\alpha_n, u_n) \in \mathcal{F}^{-1}[0]$, with $u_n \neq 0$ for all $n \geq 1$, such that

$$\lim_{n \to \infty} (\alpha_n, u_n) = (\alpha_0, 0).$$

For every map $\mathcal{F}$ satisfying (F1)–(F3), we denote

$$\mathcal{L}(\alpha) = \mathcal{D}_u \mathcal{F}(\alpha, 0), \qquad \alpha \in \mathbb{R}.$$

By property (F2), we have that $\mathcal{L} \in C(\mathbb{R}, \mathfrak{L}(U, V))$, and by (F1), we have $\mathcal{L}(\alpha) \in \text{Fred}_0(U, V)$; thus

$$\mathcal{L}(\alpha) \in \text{Iso}(U, V) \quad \Longleftrightarrow \quad \dim \text{N}[\mathcal{L}(\alpha)] = 0.$$

Lemma 2.3 ([121]) *Suppose $(\alpha_0, 0)$ is a bifurcation point of $\mathcal{F} = 0$. Then, $\alpha_0 \in \text{Sing}(\mathcal{L})$ $(\mathcal{L}(\alpha) = \mathcal{D}_u \mathcal{F}(\alpha, 0))$.*

Theorem 2.3 (Local Bifurcation [46]) *Let U and V be Banach spaces, W a neighborhood of 0 in U, and*

$$\mathcal{F} : (-1, 1) \times W \mapsto V,$$

have the properties:

1. *$\mathcal{F}[\alpha, 0] = 0$ for $|\alpha| < 1$.*
2. *The partial derivatives $\mathcal{D}_\alpha \mathcal{F}, \mathcal{D}_u \mathcal{F}, \mathcal{D}_{\alpha u} \mathcal{F}$ exist and are continuous,*
3. *$\text{N}[\mathcal{D}_u \mathcal{F}(0, 0)]$ and $V / \text{R}[\mathcal{D}_u \mathcal{F}(0, 0)]$ are one-dimensional, and we write*

$$\text{N}[\mathcal{D}_u \mathcal{F}(0, 0)] = \text{span}(u_0).$$

4. *$\mathcal{D}_{\alpha u} \mathcal{F}(0, 0)[u_0] \notin \text{R}[\mathcal{D}_u \mathcal{F}(0, 0)].$*

If Z is any complement of $\text{N}[\mathcal{D}_u \mathcal{F}(0, 0)]$ in U, then there are a neighborhood N of $(0, 0)$ in $\mathbb{R} \times U$, an interval $(-a, a)$, and continuous functions $\phi : (-a, a) \to \mathbb{R}$, $\psi : (-a, a) \to Z$ such that $\phi(0) = 0, \psi(0) = 0$ and

$$\mathcal{F}^{-1}[0] \cap N = \{ (\phi(s), \alpha u_0 + \alpha \psi(s)) : |s| < a \} \cup \{ (\alpha, 0) : (\alpha, 0) \in N \}.$$

If $\mathcal{D}_{uu} \mathcal{F}$ is continuous, then the functions ϕ and ψ are once continuously differentiable.

We have the following results on regularity, continuity, and differentiability.

Theorem 2.4 ([46]) *In addition to the assumptions of Theorem 2.3, let $\mathcal{F}$ be twice differentiable. If ϕ and ψ are the functions of Theorem 2.3, then there is $\delta > 0$ such that $\phi'(s) \neq 0$ and $0 < |s| < \delta$ implies that $\mathcal{D}_u \mathcal{F}(\phi(s), \alpha u_0 + \alpha \psi(s))$ is an isomorphism of U onto V.*

Theorem 2.5 ([46]) *In addition to the assumptions of Theorem 2.3, suppose $\mathcal{F}$ has n continuous derivatives with respect to (α, u) and $n + 1$ continuous derivatives with respect to u. Then the functions (ϕ, ψ) have n continuous derivatives with respect to s. If*

$$\mathcal{D}_u^{(j)} \mathcal{F}(0,0)[u_0]^j = 0 \qquad 1 \leq j \leq n \quad \text{then } \phi^{(j)}(0) = 0,$$

and

$$\psi^{(j)}(0) = 0 \qquad \text{for } 1 \leq j \leq n - 1,$$

and

$$\frac{1}{n+1} \mathcal{D}_u^{(n+1)} \mathcal{F}(0,0)[u_0]^{n+1} + \mathcal{D}_u \mathcal{F}(0,0)[\psi^{(n)}(0)] + \phi^{(n)}(0) \mathcal{D}_{\alpha u} \mathcal{F}(0,0)[u_0] = 0.$$

Remark 2.3 $\mathcal{D}_u^{(j)} \mathcal{F}(0,0)[u_0]^j$ means the value of the j-th Fréchet derivative of the map $x \to \mathcal{F}(0, x)$ at $(0, 0)$ evaluated at the j-tuple each of whose entries is u_0.

The global version of Theorem 2.3 reads (see Fig. 2.4).

Theorem 2.6 ([121]) *Suppose $\mathcal{L} \in C^1(\mathbb{R}, \text{Fred}_0(U, V))$ and $\alpha_0 \in \mathbb{R}$ is a simple singular point of $\mathcal{L}$, that is,*

$$N[\mathcal{L}(\alpha_0)] = \text{span}[\phi_0],$$

and satisfies the following transversality condition

$$\mathcal{L}'(\alpha_0)\phi_0 \notin R[\mathcal{L}(\alpha_0)]. \tag{2.10}$$

Then, for every continuous function $\mathcal{F} : \mathbb{R} \times U \to V$ satisfying (F1), (F2), and (F3) and $\mathcal{D}_u \mathcal{F}(\cdot, 0) = \mathcal{L}(\cdot)$, $(\alpha_0, 0)$ is a bifurcation point to a continuum $\mathfrak{C}$ of non-trivial solutions of $\mathcal{F} = 0$. For any of these $\mathcal{F}$'s, let $\mathfrak{C}$ be the component of the set of non-trivial solutions of $\mathcal{F} = 0$ with $(\alpha_0, 0) \in \mathfrak{C}$. Then either,

1. *$\mathfrak{C}$ is not compact; or*
2. *there is another Sing $\ni \alpha_1 \neq \alpha_0$ with $(\alpha_1, 0) \in \mathfrak{C}$.*

And finally we have the following unilateral result.

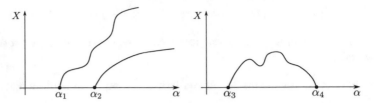

Fig. 2.4 Phase-space plot of the two possible alternatives of Theorem 2.6. On the left, the bifurcation branches are unbounded, while on the right, the non-trivial solution branch connects two bifurcation points

Theorem 2.7 ([121]) *Suppose the injection $U \hookrightarrow V$ is compact, $\mathcal{F}$ satisfies (F1)-(F3), the map*

$$N[\alpha, u] = \mathcal{F}[\alpha, u] - \mathcal{D}_u \mathcal{F}(\alpha, 0)[u], \quad (\alpha, u) \in \mathbb{R} \times U$$

admits a continuous extension to $\mathbb{R} \times V$, the transversality condition (2.10) holds, and consider a closed subspace $Y \subset U$ such that

$$U = N[\mathcal{L}(\alpha_0)] \oplus Y.$$

Let $\mathfrak{C}$ be the component given by Theorem 2.6, and denote by $\mathfrak{C}^+$ and $\mathfrak{C}^-$ the sub-components of $\mathfrak{C}$ in the directions ϕ_0 and $-\phi_0$, respectively. Then for each $v \in \{+, -\}$, $\mathfrak{C}^v$ satisfies some of the following alternatives:

1. *$\mathfrak{C}^v$ is not compact in $\mathbb{R} \times U$,*
2. *There exists $\alpha_1 \neq \alpha_0$ such that $(\alpha_1, 0) \in \mathfrak{C}^v$,*
3. *There exists $(\alpha, y) \in \mathfrak{C}^v$ with $y \in Y \setminus \{0\}$.*

In Chap. 4, we apply Theorem 2.3 to find local bifurcation branches originating from the trivial steady-state solution of the non-local equation (1.1). In Chap. 5, we apply Theorem 2.6 and Theorem 2.7 to obtain a global bifurcation result for the steady states of Eq. (1.1).

2.4 The Averaging Operator in Periodic Domains

As we are dealing with functions on a circle of length L, S_L^1, we implicitly defined periodic boundary conditions. Sometimes it is useful to explicitly use these boundary conditions. For this reason, we define the boundary operator,

$$\mathcal{B}[u, u'] := (u(0) - u(L), u'(0) - u'(L)), \tag{2.11}$$

which of course has to be equal to zero if we impose periodic boundary conditions. The abstract formulation in terms of an operator equation will be facilitated by the following operators that we will now define.

Definition 2.5 Let $L > 0$.

1. We define the set of test functions to be

$$C^\infty_{per}(0, L) := \{\, f \in C^\infty(0, L) : f \text{ is } L \text{ periodic} \,\}.$$

Then, for example, $H^1_{per}(0, L)$ is defined as the completion of $C^\infty_{per}(0, L)$ with respect to the H^1 norm.

2. The averaging operator

$$\mathcal{A} : L^p(S^1_L) \mapsto \mathbb{R}, \quad \mathcal{A}[u] \mapsto \frac{1}{L} \int_0^L u(x)\, dx, \qquad (2.12)$$

is continuous and compact.

3. We define the following sub-manifold of $L^2(S^1_L)$:

$$L^2_0 := \{\, u \in L^2(S^1_L) : \mathcal{A}[u] = 0 \,\}. \qquad (2.13)$$

Lemma 2.4 $L^2_0(S^1_L)$ *is closed and hence a Banach space.*

Proof Note that $L^2_0(S^1_L) = \mathcal{A}^{-1}(0)$; hence, it is closed as $\mathcal{A}$ is continuous. Finally, a closed subspace of a Banach space is again a Banach space. □

Lemma 2.5 *We consider the Laplace operator Δ on S^1_L.*

1. *The Laplacian operator*

$$\Delta : W^{2,2}_{per}(S^1_L) \mapsto L^2(S^1_L), \quad \Delta[v] := v''$$

is continuous.

2. *Δ is a Fredholm operator. In particular, we have that*

$$N[\Delta] = R[\mathcal{A}] = \{\, f \in H^2_p(S^1_L) : f(x) \equiv c \in \mathbb{R} \,\}, \quad R[\Delta] = N[\mathcal{A}] = L^2_0.$$

Further, we have that

$$\dim N[\Delta] = \dim L^2(S^1_L)/R[\Delta] = 1,$$

and thus $\operatorname{ind} \Delta = 0$.

3. *The restriction operator $\Delta_\mathcal{A}$*

$$\Delta_\mathcal{A} := \Delta\Big|_{N[\mathcal{A}]} : N[\mathcal{A}] \mapsto R[\Delta]$$

is an isomorphism.

Remark 2.4 Lemma 2.5 shows that $N[\Delta]$ and $\operatorname{coker}[\Delta] = L^2(S^1_L)/R[\Delta]$ are finite dimensional. We note that $\mathcal{A}$ can be used as a projection onto both $N[\Delta]$ and $\operatorname{coker}[\Delta]$.

For more detailed discussions on elliptic operators, we refer the reader to [78, 117, 169].

2.5 Local and Global Existence

The local and global existence of solutions to the adhesion model (1.1) has developed over a series of papers. The first consideration of existence and uniqueness properties is published by Sherratt et al. in [166]. In one dimension, they consider an adhesion model with a volume filling mechanism and logistic growth

$$u_t = Du_{xx} - \alpha \left(u \int_{-R}^{R} u(x+r,t) \left[2 - u(x+r,t) \right]^+ \Omega(r)\,dr \right)_x + \mu u(1-u),$$

and they show uniform boundedness of solutions by 2, assuming the solutions are classical and analytic. They extend the result to a cancer invasion model

$$u_t = Du_{xx}$$
$$- \left(u \int_{-R}^{R} (\alpha u(x+r,t) + \beta m(x+r,t)) \left[(2 - u(x+r,t) - m(x+r,t) \right]^+ \Omega(r)\,dr \right)_x$$
$$m_t = -\gamma u m^2$$

where $m(x,t)$ denotes the extracellular matrix (ECM), which is degraded by the invading tumor cell population at a rate γ.

Chaplain et al. [40] show local and global existence for a very general n-dimensional cancer-ECM adhesion model of the form

$$u_t = D\Delta u - \alpha \nabla \cdot \left(u \int_{B_R(0)} h(u,m)\Omega(r)dr \right) + u g_1(u,v)$$
$$m_t = -v g_2(m)$$
$$v_t = D_v \Delta v - \lambda v + g_3(m)u,$$

where $u(x,t)$ denotes the cell population, $m(x,t)$ the extracellular matrix and $v(x,t)$ a proteolytic enzyme (matrix metalloproteinase) that is produced by the cells, diffuses freely, and degrades the ECM. The nonlinearities g_1, g_2, and g_3 are assumed to satisfy linear growth bounds and very general Lipschitz and integrability conditions (see details in [40]). The authors show global existence of unique solutions of the corresponding initial value problem.

Unfortunately, this result does not include the general Armstrong model (1.1). The proof of global existence employs fractional differential calculus, and it is based on the regularizing properties of the MMP equation for v. The diffusion term in that equation regularizes such that m is regular, which can then be used in the first equation. If the last equation is missing, then this method no longer works.

In Winkler et al. [102], we showed global existence of the very general Armstrong model (1.1) in any space dimension. We reformulate the multidimensional adhesion model on $\mathbb{R}^n$ here:

$$u_t = D\Delta u - \alpha\nabla \cdot \left(u \int_{B_R(x)} h(u(x+r,t))\Omega(r)\,dr\right), \tag{2.14}$$

where $B_R(x)$ denotes the ball of radius $R > 0$ around x. The function $\Omega(r)$ can be written as

$$\Omega(r) = \frac{r}{|r|}\omega(|r|).$$

We assume

(A1) $h \in C^2(\mathbb{R}^n)$ and there exists a value $b > 0$ such that $h(u) = 0$ for all $u \geq b$.
(A2) $\omega \in L^1(\mathbb{R}^n)$.
(A3) For $p \geq 1$ let $u_0 \in X_p := C^0(\mathbb{R}^n) \cap L^\infty(\mathbb{R}^n) \cap L^p(\mathbb{R}^n)$ be non-negative.

Theorem 2.8 (Corollary 2.4 in [102]) *Assume (A1)–(A3). Then there exists a unique, global solution*

$$u \in C^0([0,\infty); X_p) \cap C^{2,1}(\mathbb{R}^n \times (0,\infty))$$

of (2.14) in the classical sense, with $u(0,x) = u_0(x)$, $x \in \mathbb{R}^n$.

Remark 2.5

1. Later we apply this result to the linear case of $h(u) = u$, which does not satisfy the assumption $h(u) = 0$ for $u \geq b$. However, we will show uniform global bounds for the $h(u) = u$ case, and we can simply set $h(u) = 0$ outside of the global bound for u without changing the solution.
2. The method from Winkler et al. [102] also works for the adhesion-ECM model and the adhesion-ECM-MMP model versions; see details in [102].

We see in the next section that in many cases, for example, when $h(u) = u$, the adhesion model can be written as a non-local Mckean–Vlasov equation (2.16) with adhesion potential W. This case on the n-dimensional torus $\mathbb{T}^n$ was studied by Carrillo et al. in [37], and global existence and uniqueness of classical solutions was shown in Theorem 2.2 [37], provided $W \in W^{2,\infty}(\mathbb{T}^n)$.

2.6 Adhesion Potential

We thank Paulo Amorim (Federal University of Rio de Janeiro) for a fruitful discussion on this topic.

It has been discussed in the community if the adhesion model can be written as an aggregation equation (1.3), as used in the swarming literature (see, e.g., [39, 132]). Here we show that this is indeed the case in a very general n-dimensional framework. This relation opens the doors to the rich theory of measure-valued gradient flows, and many of their results will apply here as well.

The adhesion model of Armstrong et al. [10] in n dimensions has the following form.

$$u_t = D\Delta u - \alpha\nabla \cdot \left(u \int_{B_R(0)} h(u(x+y,t))\Omega(y)\,dy \right) \qquad (2.15)$$

where the adhesion force density $h(u)$ is an increasing smooth function in u and the weight function $\Omega(y)$ satisfies:

$$\Omega(y) = \omega(\|y\|)\frac{y}{\|y\|}$$

$$\omega(\|y\|) \geq 0, \quad \omega \in L^1(B_R(0)) \cap L^\infty(B_R(0)), \quad \int_{B_R(0)^+} \omega(\|y\|)\,dy = \frac{1}{2},$$

where $B_R(0)^+$ denotes the upper half of the ball of radius R, where we can define the upper half by using any coordinate axis and consider non-negative components on this axis.

We assume that $\omega(r)$ has an antiderivative V. For physical reasons, we work with the negative of the antiderivative. That is, we assume

$$V'(r) = -\omega(r).$$

Then

$$\nabla_y V(\|y\|) = -\omega(\|y\|)\frac{y}{\|y\|} = -\Omega(y).$$

Hence $\Omega(y)$ has a potential $-V(\|y\|)$, which we call the *adhesion potential*. As the adhesion potential is unique up to addition of a constant, we can always chose V such that

$$V(R) = 0.$$

Then we can write the integral term in (2.15) as

$$\int_{B_R(0)} h(u(x+y,t))\Omega(y)\,dy = -\int_{B_R(0)} h(u(x+y,t))\nabla_y V(\|y\|)\,dy$$

$$= \int_{B_R(0)} \nabla_y h(u(x+y,t))V(\|y\|)\,dy$$

$$- \int_{\partial B_R(0)} h(u(x+y,t))V(\|y\|)\,dy$$

$$= \int_{B_R(0)} \nabla_x h(u(x+y,t))V(\|y\|)\,dy$$

$$= \nabla_x \int_{B_R(0)} h(u(x+y,t))V(\|y\|)\,dy$$

With this transformation, the adhesion equation (2.15) becomes

$$u_t = D\Delta u - \alpha \nabla \cdot \left(u \nabla \int_{B_R(0)} h(u(x + y), t) V(\|y\|)\, dy \right).$$

Finally, to write the integral term as a convolution, we use the fact that $V(R) = 0$, and we cut the potential at radius R:

$$W(r) := V(r) \chi_{[0,R]}(r).$$

Then we obtain a classical aggregation equation (1.3) with additional diffusion term

$$u_t = D\Delta u - \alpha \nabla \cdot \left[u \nabla (W * h(u)) \right]. \tag{2.16}$$

Using the known results for the aggregation equation, we then know that (2.16) is the gradient flow of the energy

$$J(u) = \int D\frac{u^2}{2}\, dx - \frac{\alpha}{2} \int u\,(W * h(u))\, dx \tag{2.17}$$

using the d_2 Wasserstein metric [6, 132].

The aggregation equation on bounded domains has recently been studied in [65, 191], and boundary conditions were defined, based on the underlying energy principle. In some cases, these boundary conditions are similar to the adhesive or repulsive boundary conditions that we discuss in Chap. 6.

For linear adhesion force $h(u) = u$, the above model (2.16) is also known as the *Mckean–Vlasov equation* (see [37]), which has manifold applications in physics, mechanics, and opinion dynamics. A full bifurcation analysis of the n-dimensional Mckean–Vlasov equation on the n-dimensional torus $\mathbb{T}^n$ was carried out recently in [37]. Their analysis is based on linearization and the properties of the free energy functional (2.17). For the case of linear adhesion force function $h(u) = u$, the results in [37], in particular Theorem 4.2, coincide with our local bifurcation result Theorem 4.13. Finally they apply the Rabinowitz alternative, to obtain a global bifurcation result. Our global bifurcation result Theorem 5.1 extends their results by providing a more detailed classification of the solution branches. While the potential W is usually defined on the entire space $\mathbb{T}^n$, Carrillo et al. present one example of a smaller sampling radius, the *noisy Hegselmann-Krause* model for opinion dynamics. In our notation, the model arises for $\omega(r) = (r - R)^+$ and $h(u) = u$.

Part II
The Periodic Problem

Chapter 3
Basic Properties

In this chapter, we define the non-local operator $\mathcal{K}[u]$, and we collect some basic properties of $\mathcal{K}[u]$ in one spatial dimension. We prove results on integrability, continuity, regularity, positivity, and a priori estimates, and we show that $\mathcal{K}[u]$ is a compact operator. We analyze the corresponding spectrum of $\mathcal{K}$, and we use these properties to derive properties of steady-state solutions such as symmetries, regularities, and a priori estimates. We find that the non-local term $\mathcal{K}[u]$ acts like a non-local derivative and the term $\mathcal{K}[u]'$ acts like a non-local curvature, in a sense made precise later.

3.1 Non-dimensionalization and Mass Conservation

We use S_L^1 to denote the circle of length L, i.e., $S_L^1 = \{\mathbb{R} \mod L\}$. The adhesion model in one dimension on the unit circle S_L^1 is given by

$$u_t(x, t) = D u_{xx}(x, t) - \alpha \left(u(x, t) \, \mathcal{K}[u(x, t)](x, t) \right)_x$$

where the operator $\mathcal{K}[u]$ is given by

$$\mathcal{K}[u(x, t)](x, t) = \int_{-R}^{R} h(u(x + r, t)) \Omega(r) \, dr.$$

To non-dimensionalize the equation, we introduce the following non-dimensional variables:

$$x^* = \frac{x}{R}, \qquad t^* = t \frac{D}{R^2}, \qquad u^* = \frac{u}{\hat{u}}, \qquad \alpha^* = \frac{\alpha}{\hat{\alpha}},$$

where $\hat{u}$ depends on the precise choice of the function $h(u)$ and $\hat{\alpha}$ is given by

$$\hat{\alpha} = \frac{D}{R\hat{u}}.$$

© The Author(s), under exclusive license to Springer Nature Switzerland AG 2021
A. Buttenschön, T. Hillen, *Non-Local Cell Adhesion Models*, CMS/CAIMS Books in Mathematics 1, https://doi.org/10.1007/978-3-030-67111-2_3

Finally let $\tilde{L} = L/R$ and $\tilde{\Omega}(\tilde{r}) = \Omega(\tilde{r}R)$ for $\tilde{r} \in [-1, 1]$. The non-dimensionalized model is then given by

$$u_t(x, t) = u_{xx}(x, t) - \alpha \left(u(x, t) \int_{-1}^{1} \tilde{h}(u(x + \tilde{r}, t))\tilde{\Omega}(\tilde{r}) \, d\tilde{r} \right)_x, \qquad (3.1)$$

for $x \in S_L^1$ and $L > 2$. In the following, to make our notation simpler, we will drop all tildes from the previous equation. Typical solutions of Eq. (3.1) are shown in Fig. 3.1.

As we do not consider any population dynamics (cell production or cell death) in Eq. (3.1), it is easy to see that mass in the system is conserved.

Lemma 3.1 *Let $u \in C^2(S_L^1)$ be a classical solution of (3.1). The total mass of the population $u(x, t)$*

$$\bar{u}(t) := \frac{1}{L} \int_0^L u(x, t) \, dx, \qquad (3.2)$$

is conserved over time.

Proof We proceed by computing

$$L\frac{d\bar{u}}{dt} = \int_0^L u_t(x, t) \, dx = \int_0^L (u_x - \alpha u \, \mathcal{K}[u])_x \, dx$$
$$= u_x(L) - u_x(0) - \alpha u(0) \, (\mathcal{K}[u](L) - \mathcal{K}[u](0)) = 0.$$

3.2 The Non-local Operator in 1-D

The key part of the adhesion model is the non-local adhesion operator as defined as follows.

Definition 3.1 Let X and Y be Banach spaces of functions; then we define the operator $\mathcal{K} : X \to Y$ by

$$\mathcal{K}[u(x)](x) = \int_{-1}^{1} h(u(x + r))\Omega(r) \, dr. \qquad (3.3)$$

The domain of integration $[x - 1, x + 1]$ is called the *sensing domain*.

Notice that we normalized the sensing radius to $R = 1$ and we assume that the domain is larger than one sensing diameter, i.e., $L > 2$. The directionality function Ω is assumed to satisfy the following conditions:

(K1) $\Omega(r) = \frac{r}{|r|}\omega(r)$, where $\omega(r)$ is an even function.
(K2) $\omega(r) \geq 0$,
(K3) $\omega \in L^1(0, 1) \cap L^\infty(0, 1)$,
(K4) $\|\omega\|_{L^1(0,1)} = 1/2$,

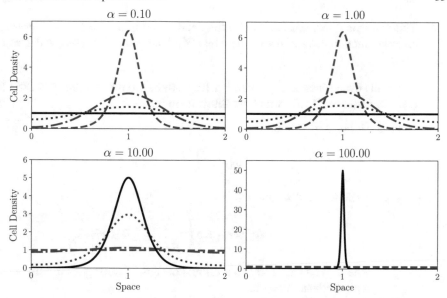

Fig. 3.1 Typical solution behavior of Eq. (3.1) for varying values of α. The initial condition is shown in blue (dashed), while the final steady-state solution is shown in black (solid). The remaining curves show intermediate times

(K5) $M_n(\omega) > 0$ for infinitely many integers n, where

$$M_n(\omega) = \int_0^1 \sin\left(\frac{2\pi nr}{L}\right) \omega(r)\, dr. \tag{3.4}$$

Condition (K1) and condition (K3) imply that

$$\int_{-1}^1 \Omega(r)\, dr = 0 \qquad \text{and} \qquad \int_{-1}^1 \omega(|r|)\, dr = 1.$$

The Fourier sine coefficients $M_n(\omega)$ of ω in (K5) are related to the eigenvalues of $\mathcal{K}$, as we will show later. The function $h(\cdot)$ within the integral describes the nature of the adhesive force and is assumed to satisfy:

(H1) $h \in C^2(\mathbb{R})$,
(H2) $h(u) \geq 0$ for $u \geq 0$,
(H3) $h(u) \leq C(1 + u)$ for all $u \geq 0$, for some $\mathbb{R} \ni C \geq 0$,
(H4) There exist positive real numbers $\bar{u}$ such that $h'(\bar{u}) \neq 0$.

Remark 3.1 Using the assumptions (K1) to (K3), we can rewrite the non-local function defined in Eq. (3.3) as

$$\mathcal{K}[u] : x \mapsto \int_0^1 [\, h(u(x + r)) - h(u(x - r))\,]\, \omega(r)\, dr.$$

This equivalent formulation will be frequently used in the following. Since $\omega(r)$ is an even function, we see already here that $\mathcal{K}[u]$ appears as a non-local derivative of $h(u(x))$.

For the even function $\omega(r)$ with $r \in [0, 1]$, there are three commonly used forms (see, for instance, [146]), which we illustrate in Fig. 3.2.

(O1) Uniform distribution

$$\omega(r) = \frac{1}{2}, \qquad 0 \le r \le 1.$$

(O2) Exponential distribution

$$\omega(r) = \omega_0 \exp\left(-\frac{r}{\xi}\right), \qquad 0 \le r \le 1,$$

where ξ is a parameter controlling how quickly $\omega(\cdot)$ goes to zero and ω_0 is a normalization constant.

(O3) Peak signalling a distance ξ away from the cell center.

$$\omega(r) = \omega_0 \frac{r}{\xi} \exp\left(-\frac{1}{2}\left(\frac{r}{\xi}\right)^2\right), \qquad 0 \le r \le 1,$$

where ξ is a parameter controlling how quickly $\omega(\cdot)$ goes to zero and ω_0 is a normalization constant.

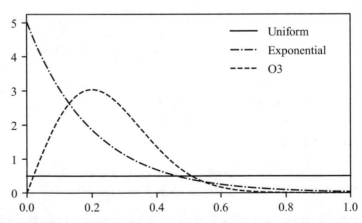

Fig. 3.2 The different distributions for $\omega(\cdot)$ with $\xi = 1/4$

The next lemma provides an example of a situation in which there are integers n such that $M_n(\omega) = 0$. Suppose that $\omega \equiv 1/2$ and that $L = 2kR$ where $k \in \mathbb{N}$. Then we have the following result, indicating a degenerate situation.

Lemma 3.2 *Let $L = 2k$, for $\mathbb{N} \ni k > 0$, and $\omega(r) \equiv {}^1/_2$; then for $n \in \mathbb{N}$ such that $\frac{n}{k}$ is an even integer, we have that*

$$M_n(\omega) = 0.$$

Proof The term $M_n(\omega)$ is defined in Eq. (3.4) as

$$M_n(\omega) = \int_0^1 \sin\left(\frac{2\pi nx}{L}\right) \Omega(r)\, dr = \frac{1}{2} \int_0^1 \sin\left(\frac{\pi nx}{k}\right) dr,$$

and then this is zero whenever $\frac{n}{k}$ is an even integer.   □

The significance of $M_n(\omega)$ being zero is that under those circumstances we do not have a bifurcation point (see Chap. 4). As we will show later, if $M_n(\omega)$ is zero, it is impossible to obtain a steady-state solution having n-peaks. Further note that both the properties of ω and the domain length L determine whether $M_n(\omega)$ is zero.

In the next lemma, we establish continuity and L^p properties of the operator $\mathcal{K}[u]$.

Lemma 3.3 *Assume (K1–K4) and (H1–H4). We assume $p \geq 1$. The function*

$$\mathcal{K}[u] : x \mapsto \mathcal{K}[u](x), \tag{3.5}$$

has the following properties:

(P1) *Let $\mathbb{N} \ni s \leq 2$. If $u \in C^s(S_L^1)$, then $\mathcal{K}[u] \in C^s(S_L^1)$.*
(P2) *If $u \in L^p(S_L^1)$, then $\mathcal{K}[u] \in C^0(S_L^1)$.*
(P3) *If $u \in L^p(S_L^1)$, then $\mathcal{K}[u] \in L^p(S_L^1)$, and there exists a constant $\tilde{C} > 0$ such that*

$$|\mathcal{K}[u]|_p \leq |h(u)|_p \leq \tilde{C}\left(|u|_p + L\right).$$

(P4) *Assume $k = 1, 2$. If $u \in W^{k,p}$, then $\mathcal{K}[u] \in W^{k,p}$.*
(P5) *The operator $\mathcal{K} : L^p \to L^p$ is continuously Fréchet differentiable in u.*
(P6) *If $u \in L^p(S_L^1)$ such that $u(x) \geq 0$ (i.e., $|u|_1 = L$, $\mathcal{A}[u] < \infty$), then*

$$|\mathcal{K}[u](x)| \leq C|\omega|_\infty L\left(\mathcal{A}[u] + 1\right), \quad \text{with} \quad \mathcal{A}[u] = \frac{1}{L}\int_0^L u(x)\, dx.$$

Proof

1. If $u \in C^0(S_L^1)$, then

$$\mathcal{K}[u](x) = \int_{-1}^1 h(u(x + r))\Omega(r)\, dr$$

is continuous in x, since h is continuous. If $u \in C^1(S_L^1)$, then

$$\mathcal{K}[u](x)' = \int_{-1}^1 h'(u(x + r))u'(x + r)\Omega(r)\, dr$$

is continuous in x, since h is differentiable. If $u \in C^2(S_L^1)$, then

$$\mathcal{K}[u](x)'' = \int_{-1}^{1} \left(h''(u(x+r))(u'(x+r))^2 + h'(u(x+r))u''(x+r) \right) \Omega(r) \, dr$$

is continuous in x, since h is C^2.

2. Let $u \in L^p(S_L^1)$. Let $\epsilon > 0$ and let $x_1, x_2 \in S_L^1$ such that $|x_1 - x_2| < \delta$. By the density of C^0 in L^p, there is a sequence $(u_n) \subset C^0$ such that $u_n \to u$ in L^p. This means that $\exists N : \forall n \geq N$; we have that $|u_n - u|_p < \epsilon/3|\omega|_\infty$. Then, we compute

$$|\mathcal{K}[u](x_1) - \mathcal{K}[u](x_2)| \leq |\mathcal{K}[u](x_2) - \mathcal{K}[u_n](x_2)| + |\mathcal{K}[u_n](x_1) - \mathcal{K}[u](x_1)| + |\mathcal{K}[u_n](x_2) - \mathcal{K}[u_n](x_1)|.$$

The first two terms are treated equivalently. Let $n > N$ and $i = 1, 2$; then

$$\begin{aligned}
|\mathcal{K}[u](x_i) - \mathcal{K}[u_n](x_i)| &\leq \int_0^1 |h(u(x_i + r)) - h(u_n(x_i + r))|\omega(r) \, dr \\
&+ \int_0^1 |h(u(x_i - r)) - h(u_n(x_i - r))|\omega(r) \, dr \\
&\leq C \int_0^1 |u(x_i + r) - u_n(x_i + r)|\omega(r) \, dr \\
&+ C \int_0^1 |u(x_i - r) - u_n(x_i - r)|\omega(r) \, dr \\
&\leq \left(\int_0^L |u(x) - u_n(x)|^p \, dx \right)^{1/p} \left(\int_0^1 \omega^q(r) \, dr \right)^{1/q} \\
&\leq |u - u_n|_p |\omega|_\infty < \epsilon/3.
\end{aligned}$$

The last term can be estimated by continuity of $\mathcal{K}[u_n](x)$ as

$$|\mathcal{K}[u_n](x_2) - \mathcal{K}[u_n](x_1)| < \epsilon/3.$$

Putting everything together, we obtain

$$|\mathcal{K}[u](x_1) - \mathcal{K}[u](x_2)| < \epsilon.$$

This shows that $\mathcal{K}[u](x) \in C^0(S_L^1)$.

3. Let $1 \leq p$, and let $u \in L^p(S_L^1)$. By applying the Minkowski integral inequality to Eq. (3.3), we obtain

$$|\mathcal{K}[u](x)|_p = \left| \int_{-1}^{1} h(u(x+r,t))\Omega(r)\,dr \right|_p \le \int_{-1}^{1} |h(u(x+r,t))\Omega(r)|_p \,dr$$

$$= \int_{-1}^{1} \left\{ \int_{0}^{L} |h(u(x+r,t))|^p \,dx \right\}^{1/p} |\Omega(r)|\,dr \le |h(u)|_p$$

$$\le \tilde{C}\left(|u|_p + L\right),$$

using assumption (H3). Note that due to assumption (K3), we have that $|\Omega|_1 = 1$.

4. Let $u \in W^{1,p}(S_L^1)$. The first derivative with respect to x of the function defined in Eq. (3.5) is given by

$$(\mathcal{K}[u](x))' = \int_{-1}^{1} h'(u(x+r))u'(x+r)\Omega(r)\,dr.$$

Its L^p norm can be estimated using (P3); we then compute

$$\|\mathcal{K}[u]\|_{1,p} = \left(|\mathcal{K}[u]|_p^p + |(\mathcal{K}[u])'|_p^p\right)^{1/p} \le \left(\tilde{C}^p\left(|u|_p + L\right)^p + |h'(u)u'|_p^p\right)^{1/p}$$

$$\le \left(2^p \tilde{C}^p\left(L^p + |u|_p^p\right) + |h'|_{C^0}^p |u'|_p^p\right)^{1/p}$$

$$\le 2\tilde{C}|h'|_{C^0}\left(L^p + |u|_p^p + |u'|_p^p\right)^{1/p}.$$

Then using that $u \in W^{1,p}(S_L^1)$ and assumption (H1), all the terms on the right-hand side are bounded.

Now let $u \in W^{2,p}(S_L^1)$. The second derivative with respect to x of the function defined in Eq. (3.5) is given by

$$(\mathcal{K}[u](x))'' = \int_{-1}^{1} \left(h'(u(x+r))u''(x+r) + h''(u(x+r))\left(u'(x+r)\right)^2\right)\Omega(r)\,dr.$$

Its L^p norm can be estimated using (P3); we then compute

$$|\mathcal{K}[u]''|_p \le |h'(u)u''|_p + |h''(u)\left(u'\right)^2|_p$$

$$\le |h'|_{C^0}|u''|_p + |h''|_{C^0}|u'|_p^2.$$

Then using that $u \in W^{2,p}(S_L^1)$ and assumption (H1), all the terms on the right-hand side are bounded. Combining this result with the estimates of (P4), we obtain the required result.

5. Consider the map

$$\mathcal{K} : L^p(S_L^1) \mapsto L^p(S_L^1).$$

The Fréchet derivative of $\mathcal{K}$ is computed by

$$\mathcal{D}_u\left(\mathcal{K}[u]\right)[w](x) = \frac{d}{d\epsilon}\Big|_{\epsilon=0} \int_{-1}^{1} h(u(x+r) + \epsilon w(x+r))\Omega(r)\,dr$$

$$= \int_{-1}^{1} h'(u(x+r))w(x+r)\Omega(r)\,dr.$$

We show that the map

$$\mathcal{D}_u\,\mathcal{K} : L^p(S_L^1) \mapsto \mathfrak{L}(L^p(S_L^1), L^p(S_L^1))$$
$$u \mapsto d_u\left(\mathcal{K}[u]\right)[w],$$

is continuous. Meaning that we have to show that the operator norm of $\mathcal{D}_u\,\mathcal{K}$ is bounded. From (P3), we have

$$|\mathcal{D}_u\left(\mathcal{K}[u]\right)[w]|_p \leq \tilde{C}\left(|h'|_{C^0}|w|_p + L\right).$$

Hence

$$\|\mathcal{D}_u\left(\mathcal{K}[u]\right)[w]\|_{op} = \sup_{|w|_p=1} |\mathcal{D}_u\left(\mathcal{K}[u]\right)[w]|_p \leq \tilde{C}\left(|h'|_{C^0} + L\right).$$

6. Let $1 \leq p$ such that $\mathcal{A}[u] < \infty$ and $u(x) \geq 0$. Then, we compute

$$\mathcal{K}[u](x) = \int_{-1}^{1} h(u(x+r))\Omega(r)\,dr$$

$$= \int_{0}^{1} h(u(x+r))\omega(r)\,dr - \int_{-1}^{0} h(u(x+r))\omega(r)\,dr.$$

It is easy to see that both integrals on the right-hand side are non-negative as $h(u(x)) \geq 0$ whenever $u(x) \geq 0$ (assumption (H2)) and $\omega(r) \geq 0$. Using assumption (H3), it follows that

$$\mathcal{K}[u](x) \leq \int_{0}^{1} h(u(x+r))\omega(r)\,dr \leq C|\omega|_\infty L\left(\mathcal{A}[u] + 1\right).$$

In the same spirit, we find for

$$\mathcal{K}[u](x) \geq -\int_{-1}^{0} h(u(x+r))\omega(r)\,dr \geq -C|\omega|_\infty L\left(\mathcal{A}[u] + 1\right).$$

Lemma 3.4 *The operator* $\mathcal{K} : L^2(S_L^1) \mapsto L^2(S_L^1)$ *is compact.*

Proof Since $L^2(S_L^1)$ is a Hilbert space, we can use an orthonormal system $\{\phi_i\}_{i\in\mathbb{N}}$ to define a finite truncation of the integral operator. We denote the inner product in L^2 with $(\cdot, \cdot)$. Given $u \in L^2(S_L^1)$, we consider the expression $h(u(x+r))$ to be a function of two variables x and r, and we expand it as

$$h(u(x + r)) = \sum_{i,j=1}^{\infty} h_{ij}(u)\phi_i(x)\phi_j(r)$$

with

$$h_{ij}(u) = \Big(h(u(x + r)), \phi_i(x)\phi_j(r)\Big).$$

We define a finite approximation of $h(u)$ as

$$\Big(h(u(x + r))\Big)_n := \sum_{i,j=1}^{n} h_{ij}(u)\phi_i(x)\phi_j(r)$$

and the corresponding truncated integral operator

$$\mathcal{K}_n[u](x) := \int_{-1}^{1} \Big(h(u(x + r))\Big)_n \Omega(r)\,dr$$

$$= \sum_{i,j=1}^{n} \int_{-1}^{1} h_{i,j}(u)\phi_j(r)dr\,\phi_i(x)$$

$$\in \text{span}\{\phi_1, \cdots, \phi_n\}.$$

For each $n > 0$, the operator $\mathcal{K}_n$ has finite rank; hence, it is compact ([160, Lemma 3.12]). If we can show that $\mathcal{K}_n \to \mathcal{K}$ as $n \to \infty$ in the operator norm, then by [160, Theorem 3.10] also $\mathcal{K}$ is compact. Indeed

$$\|\mathcal{K} - \mathcal{K}_n\|_{op} = \sup_{\|u\|=1} |\mathcal{K}u - \mathcal{K}_n u|_2^2$$

with

$$|\mathcal{K} - \mathcal{K}_n|_2^2 \le \int_0^L \int_{-1}^1 \Big| h(u(x + r)) - \Big(h(u(x + r))\Big)_n \Big|^2 \Omega^2(r)\,dr\,dx$$

$$\le \int_0^L \int_0^L \Big| \sum_{i,j=n+1}^{\infty} h_{ij}(u)\phi_i(x)\phi_j(r) \Big|^2 \Omega^2(r)\,dr\,dx$$

$$\le \sum_{i,j=n+1}^{\infty} |h_{ij}(u)|^2 \|\Omega\|_\infty^2$$

$$\to 0 \quad \text{as} \quad n \to \infty. \qquad\qquad \square$$

The compactness of $\mathcal{K}$ has far-reaching consequences, namely, it means that $\mathcal{K}[u]$ can never be invertible, as otherwise $\mathcal{I} = \mathcal{K}^{-1}\mathcal{K}$ would be compact. Further a compact operator can never be subjective, and the closed subspaces of its range must be finite dimensional. Also, the spectral theorems for compact operators apply to $\mathcal{K}$, which we will employ next.

3.3 Spectral Properties

In this section, we consider the spectral properties of linear non-local operator $\mathcal{K}[u]$ (i.e., $h(u) = u$) and of the non-local curvature $\mathcal{K}[u]'$. That is,

$$\mathcal{K}[u] = \int_{-1}^{1} u(x + r)\Omega(r)\,dr. \tag{3.6}$$

The results of this section will be used in Chap. 4 to study the properties of the linearization of the steady-state equation. We start by a result how $\mathcal{K}[u]$ acts on basis functions of L^2.

Lemma 3.5 *Let Ω satisfy (K1), (K2), and (K3); then*

$$\int_{-1}^{1} \cos\left(\frac{2\pi nr}{L}\right)\Omega(r)\,dr = 0,$$

and

$$\int_{-1}^{1} \sin\left(\frac{2\pi nr}{L}\right)\Omega(r)\,dr = 2\int_{0}^{1} \sin\left(\frac{2\pi nr}{L}\right)\omega(r)\,dr.$$

Proof Notice that $\Omega(r) = \frac{r}{|r|}\omega(r)$, where $\omega(r)$ is an even function; hence, both identities follow by integration and symmetry.   □

Lemma 3.6 *Consider the operator $\mathcal{K}[u]$ with $h(u) = u$ as defined in Eq. (3.6) as an operator*

$$\mathcal{K} : L^2(S_L^1) \mapsto L^2(S_L^1).$$

$\mathcal{K}$ has the following properties:

1. *$\mathcal{K}$ is bounded and skew-adjoint, i.e., $\mathcal{K}^* = -\mathcal{K}$.*
2. *$\mathcal{K}$ maps the canonical basis functions of $L^2(S_L^1)$ as follows:*

$$\mathcal{K}[1](x) = 0,$$

$$\mathcal{K}\left[\sin\left(\frac{2\pi nx}{L}\right)\right](x) = 2M_n(\omega)\cos\left(\frac{2\pi nx}{L}\right),$$

$$\mathcal{K}\left[\cos\left(\frac{2\pi nx}{L}\right)\right](x) = -2M_n(\omega)\sin\left(\frac{2\pi nx}{L}\right),$$

where $M_n(\omega)$ is defined in Eq. (3.4).
3. *If instead of $L^2(S_L^1)$ we consider its canonical complexification, then we find complex eigenvalues:*

$$\mathcal{K}[1](x) = 0,$$

$$\mathcal{K}\left[\exp\left(\frac{2\pi nix}{L}\right)\right](x) = 2iM_n(\omega)\exp\left(\frac{2\pi nix}{L}\right).$$

Proof $\mathcal{K}$ is bounded by Lemma 3.3. To see the skew-adjointness, we consider for $u, v \in L^2(S_L^1)$.

$$
\begin{aligned}
(\mathcal{K}[u], v) &= \int_0^L \int_{-1}^1 u(x + r)\Omega(r)\,dr\,v(x)\,dx \\
&= \int_{-1}^1 \int_r^{L+r} u(y)\Omega(r)v(y - r)\,dr\,dy \\
&= \int_0^L \int_{-1}^1 u(y)\Omega(r)v(y - r)\,dr\,dy \\
&= -\int_0^L \int_{-1}^1 v(y + r)\Omega(r)\,dr\,u(y)\,dy \\
&= -(u, \mathcal{K}[v]),
\end{aligned}
$$

where we used periodicity in the second step and the property that Ω is odd in the third step. Properties (2) and (3) follow by direct computation, using the symmetries of sin, cos, and Ω. $\square$

Remark 3.2

1. Note that in general skew-adjoint operators have purely imaginary eigenvalues.
2. Note that since any skew-adjoint operator is normal, we have that $\mathcal{K}$ is a normal operator. Thus we have that $\mathcal{K}$ is a compact and normal operator on $L^2(S_L^1)$. This means that it is also a compact and normal operator on the canonical complexification of $L^2(S_L^1)$ ($H = L^2 + iL^2$) and hence we can apply a spectral theorem [63, Theorem 7.53] to obtain an orthonormal basis of H over which the operator $\mathcal{K}$ is diagonalizable.
3. Note that it is easy to see from Lemma 3.6 that the non-local operator $\mathcal{K}[u]$ removes mass, in the sense that $\mathcal{A}[\mathcal{K}[u]] = 0$ (the average operator $\mathcal{A}$ was defined in (2.12)). Hence, for example, if $\mathcal{K} : L^2(S_L^1) \to L^2(S_L^1)$, then we conclude that $R[\mathcal{K}] = L_0^2$ (where L_0^2 contains L^2 functions with mass zero, as defined in (2.13)).
4. For the sake of comparison, the eigenvalues corresponding to the eigenfunctions $\exp\left(\frac{2\pi n i x}{L}\right)$ of the derivative operator are

$$
\lambda_n = \frac{2\pi i n}{L}.
$$

Let us consider the spectrum of the derivative of $\mathcal{K}[u]$. From Lemma 3.3 (P4), we know that $\mathcal{K}[u]'$ maps $H^1(S_L^1)$ into $H^1(S_L^1)$. Here we study the operator, which we will refer to as the linear *non-local curvature*

$$
\mathcal{K}[u]' : x \mapsto \left(\int_{-1}^1 u(x + r)\Omega(r)\,dr\right)', \tag{3.7}
$$

where $(\cdot)'$ denotes the spatial derivative with respect to x. Note that using the properties of Ω, this function can be rewritten as

$$(\mathcal{K}[u])' : x \mapsto \int_0^1 (u'(x+r) + u'(x-r))\,\omega(r)\,dr.$$

Lemma 3.7 *The operator* $(\mathcal{K})'$ *given above is self-adjoint.*

Proof Let $y, z \in L^2(S_L^1)$; we then compute using integration by parts and using Lemma 3.6 to obtain

$$((\mathcal{K}[y])', z) = -(\mathcal{K}[y], z') = (y, \mathcal{K}[z']) = (y, (\mathcal{K}[z])').$$

Lemma 3.8 *Let* v_n *be an eigenfunction of*

$$\begin{cases} -v_n'' = \lambda_n v_n, & in \; [0, L] \\ \mathcal{B}[v_n, v_n'] = 0, \end{cases}$$

that is,

$$v_n \in \left\{ 1, \sin\left(\frac{2n\pi x}{L}\right), \cos\left(\frac{2n\pi x}{L}\right), \; n = 1, 2, \dots \right\}$$

Then the operator $(\mathcal{K})'$ *from Eq. (3.7) has the same set of eigenfunctions satisfying*

$$\mathcal{K}[v_n]' = \mu_n v_n,$$

where

$$\mu_n = -\frac{4\pi n}{L} M_n(\omega),$$

where $M_n(\omega)$ *is defined in Eq. (3.4).*

Proof For $n = 0$, we have that $v_0 = 1$, then trivially $\mathcal{K}[1]' = 0$, and hence $\mu_0 = 0$. Next consider $v_n = \sin\left(\frac{2\pi n x}{L}\right)$; then

$$\mathcal{K}[v_n]' = \int_{-1}^1 \left(\sin\left(\frac{2\pi n(x+r)}{L}\right)\right)' \Omega(r)\,dr.$$

Using the symmetries of sin, cos, and Ω, we obtain

$$\mathcal{K}[v_n]' = -\frac{4\pi n}{L} \sin\left(\frac{2\pi n x}{L}\right) M_n(\omega).$$

Finally, consider $v_n = \cos\left(\frac{2\pi n x}{L}\right)$; then

$$\mathcal{K}[v_n]' = \int_{-1}^1 \left(\cos\left(\frac{2\pi n(x+r)}{L}\right)\right)' \Omega(r)\,dr.$$

Once again using symmetries, we obtain

$$\mathcal{K}[v_n]' = -\frac{4\pi n}{L} \cos\left(\frac{2\pi nx}{L}\right) M_n(\omega).$$

From the definition of $M_n(\omega)$ in Eq. (3.4), we easily see from (K4) that $|M_n(\omega)| < 1/2$.

Here, we want to understand in more detail how $M_n(\omega)$ behaves as $n \to \infty$. For this reason, we introduce the following common definition from Fourier analysis (see, for instance, [194, Chapter II]).

Definition 3.2 The integral *modulus of continuity* is defined for periodic $f \in L^p(S_L^1)$, $p \geq 1$ by

$$m_p(\delta) = \sup_{0 \leq h \leq \delta} \left\{ \frac{1}{L} \int_0^L |f(x+h) - f(x)|^p \, dx \right\}^{1/p}, \qquad \delta > 0.$$

It is obvious that as $\delta \to 0$, we have that $m_p(\delta) \to 0$.

Lemma 3.9 *Let $\omega(r)$ satisfy (K1), (K2), and (K3) and $M_n(\omega)$ be given by Eq. (3.4)); then we have that*

$$|M_n(\omega)| \leq \frac{L}{2} m_1 \left(\frac{L}{2n}\right).$$

Proof First, extend $\omega(r)$ to the whole of $[0, L]$ by defining the L-periodic function

$$\tilde{\omega}(r) = \begin{cases} \omega(r) & \text{if } r \leq 1 \\ 0 & \text{otherwise} \end{cases}.$$

$\square$

Then, we use a common technique from Fourier theory (see, for instance, [194, Chapter II]).

$$M_n(\omega) = \int_0^L \sin\left(\frac{2\pi nr}{L}\right) \tilde{\omega}(r) \, dr$$

$$= -\int_0^L \sin\left(\frac{2\pi n}{L}\left(r - \frac{L}{2n}\right)\right) \tilde{\omega}(r) \, dr$$

$$= -\int_0^L \sin\left(\frac{2\pi nr}{L}\right) \tilde{\omega}\left(r + \frac{L}{2n}\right) \, dr.$$

Taking the average of both integrals, we obtain

$$M_n(\omega) = \frac{1}{2} \int_0^L \left(\tilde{\omega}(r) - \tilde{\omega}\left(r + \frac{L}{2n}\right)\right) \sin\left(\frac{2\pi nr}{L}\right) \, dr.$$

Hence,

$$M_n(\omega) \leq \frac{L}{2} \sup_{0 < h \leq \frac{L}{2n}} \frac{1}{L} \int_0^L |\tilde{\omega}(r) - \tilde{\omega}(r+h)| \, dr = \frac{L}{2} m_1\left(\frac{L}{2n}\right).$$

Example 3.1 Suppose that $\omega(r) = \frac{1}{2}$ is chosen to be the uniform function, example, i.e., (O1). Then, we can compute $M_n(\omega)$ and find that

$$M_n(\omega) = \frac{L}{2\pi n} \sin^2\left(\frac{\pi n}{L}\right).$$

Hence, $M_n(\omega) \to 0$ as $n \to \infty$, and thus so do the eigenvalues of $\mathcal{K}$ from Eq. (3.6) (see Lemma 3.6) (since $\mathcal{K}$ is compact, this is expected as the only possible accumulation point of the eigenvalues is zero). But the eigenvalues of the non-local curvature (3.7) are given by (see Lemma 3.8)

$$\mu_n = -\frac{4\pi n}{L} M_n(\omega) = -2\sin^2\left(\frac{\pi n}{L}\right).$$

Thus, the eigenvalues of the non-local curvature keep oscillating in $(-2, 0)$.

3.4 The Behavior of $\mathcal{K}_R$ for $R \to 0$

In this subsection, we show that the linear non-local operator $\mathcal{K}[u]$ defined in Definition 3.1 is related to the classical derivative. In particular, we show that for smooth functions h and u, the non-local operator approximates the gradient as the sensing radius R converges to zero; hence, we call it a non-local gradient. For this section, we give up the normalization of a sensing radius of $R = 1$, and we consider the non-local operator for general $R > 0$

$$\mathcal{K}_R[u] := \int_{-R}^{R} h(u(x + r))\Omega(r)\,dr,$$

where $\Omega(r)$ is defined by

$$\Omega(r) = \frac{r}{|r|}\frac{\omega(r/R)}{R},$$

with $\omega(\cdot)$ being the function introduced in Definition 3.1. In this case

$$\int_0^R \Omega(r)\,dr = \int_0^R \frac{r}{|r|}\frac{\omega(r/R)}{R}\,dr = \int_0^1 \omega(\sigma)\,d\sigma = \frac{1}{2}.$$

Our aim is to develop an asymptotic expansion $\mathcal{K}_R[u]$ as $R \to 0$. In preparation, we compute the moments of the distribution Ω.

$$\mu_n = \int_{-R}^{R} r^n \Omega(r)\,dr = \begin{cases} 0 & \text{if } n \text{ even} \\ 2R^n \int_0^1 \sigma^n \omega(\sigma)\,d\sigma & \text{if } n \text{ odd} \end{cases}.$$

In case where R is small, we can consider a Taylor expansion of the integrand.

$$\mathcal{K}_R[u] = \int_{-R}^{R} h(u(x+r))\Omega(r)\,dr$$

$$\approx \int_{-R}^{R} \left(h(u(x)) + r\frac{\partial}{\partial x}h(u(x)) + \frac{r^2}{2}\frac{\partial^2}{\partial x^2}h(u(x)) + \frac{r^3}{6}\frac{\partial^3}{\partial x^3}h(u(x)) + \ldots \right)\Omega(r)\,dr$$

$$= \sum_{j=0}^{\infty} \frac{\mu_j r^j}{j!}\frac{\partial^j}{\partial x^j}h(u(x))$$

$$= \mu_1\frac{\partial}{\partial x}h(u(x)) + \frac{\mu_3}{6}\frac{\partial^3}{\partial x^3}h(u(x)) + \ldots.$$

Hence, $\mathcal{K}_R[u]$ for small R acts like a derivative. It should be noted that the moments μ_j scale as R^j for j odd; hence, for differentiable functions h and u, we find

$$\frac{\mathcal{K}_R[u]}{2R} \to c_1\frac{\partial}{\partial x}h(u(x)), \quad \text{for} \quad R \to 0,$$

with

$$c_1 = \int_0^1 \sigma\omega(\sigma)\,d\sigma.$$

We illustrate this relationship for a given test function in Fig. 3.3, where we plot $c_1 u'(x)$ and $\mathcal{K}_R[u](x)$ for several values of R. We can obtain a similar asymptotic expansion for the non-local curvature (i.e., the derivative of $\mathcal{K}_R[u]$ as $R \to 0$)

$$\frac{\mathcal{K}_R[u](x)'}{4R^2} \approx \frac{\partial^2}{\partial x^2}h(u(x)) + O(R^2).$$

Example 3.2 Suppose that $\omega(\sigma) = 1/2$ for $\sigma \in [0,1]$. Then $\Omega(r) = \frac{1}{2R}\frac{r}{|r|}$ and

$$\mathcal{K}_R[u](x) = \frac{1}{2R}\int_{-R}^{R} h(u(x+r))\frac{r}{|r|}\,dr \approx \frac{R}{2}\frac{\partial}{\partial x}h(u(x)),$$

for small R.

Example 3.3 Consider a singular measure ω of the form

$$\omega(r) = \delta_R(r); \quad \text{then} \quad \Omega(r) = \delta_R(r) - \delta_{-R}(r).$$

Then

$$\mathcal{K}_R[u] = \frac{1}{2}[\,h(u(x+R)) - h(u(x-R))\,],$$

where we used the convention that a δ-distribution on the domain boundary carries weight $1/2$.

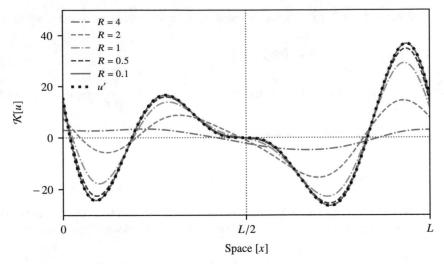

Fig. 3.3 Comparison of the non-local term to the first derivative of a function for several values for the sensing radius R

3.5 Properties of Steady-State Solutions

In this section, we prove several properties of solutions of the steady-state equation

$$u''(x) = \alpha \Big(u(x) \, \mathcal{K}[u](x) \Big)', \quad x \in S_L^1. \tag{3.8}$$

These will be useful later when we carry out the bifurcation analysis. Throughout this section, we assume (K1–K5) and (H1–H4).

Lemma 3.10 *Let $u \in L^2(S_L^1)$ be L-periodic and the nonlinearity within the non-local term be linear, i.e., ($h(u) = u$); then $\mathcal{A}[u(x)\,\mathcal{K}[u](x)] = 0$, where $\mathcal{A}[\cdot]$ is the averaging operator defined in Eq. (2.12).*

Proof We proceed by simple calculation and an application of Fubini's theorem.

$$\int_0^L u(x)\,\mathcal{K}[u](x)\,\mathrm{d}x = \int_0^L u(x) \int_{-1}^1 u(x+r)\Omega(r)\,\mathrm{d}r\,\mathrm{d}x$$

$$= \int_0^L u(x) \int_0^1 (u(x+r) - u(x-r))\,\omega(r)\,\mathrm{d}r\,\mathrm{d}x$$

$$= \int_0^1 \int_0^L u(x)\,(u(x+r) - u(x-r))\,\mathrm{d}x\,\omega(r)\,\mathrm{d}r.$$

Since u is periodic, we have

$$\int_0^L u(x)u(x+r)\,dx = \int_0^L u(x)u(x-r)\,dx,$$

for all $r \in [0, 1]$, and the result follows. □

Note that in the proof of Lemma 3.10, we required that the nonlinearity within the non-local term is the simple linear function $(h(u) = u)$. The previous proof does not work for nonlinear $h(u)$. We can however recover the result by imposing an additional symmetry assumption.

Lemma 3.11 *Let $u \in L^2(S_L^1)$ be L-periodic and $u(x) = u(L - x)$; then the average $\mathcal{A}[u(x)\,\mathcal{K}[u](x)] = 0$, where $\mathcal{A}[\cdot]$ is the averaging operator defined in Eq. (2.12) and $h(\cdot)$ satisfies H1–H4.*

Proof Due to the fact that $u(x) = u(L - x)$, we obtain that

$$u(x)\,\mathcal{K}[u](x) = -u(L - x)\,\mathcal{K}[u](L - x).$$

Hence, upon integration, we obtain

$$\int_0^L u(x)\,\mathcal{K}[u](x)\,dx = \int_0^{L/2} u(x)\,\mathcal{K}[u](x)\,dx + \int_{L/2}^L u(x)\,\mathcal{K}[u](x)\,dx$$

$$= \int_0^{L/2} u(x)\,\mathcal{K}[u](x)\,dx - \int_0^{L/2} u(x)\,\mathcal{K}[u](x)\,dx = 0.$$

Remark 3.3 Note that if $u \in H^2$, then Lemma 3.10 or 3.11 implies that the flux defined by

$$J(x) = u'(x) - \alpha u(x)\,\mathcal{K}[u(x)](x)$$

satisfies $\mathcal{A}[J] = 0$, and since J is continuous, we have that $\exists \hat{x} \in S_L^1 : J(\hat{x}) = 0$. Note that the continuity of J follows as $u \in H^2 \subset\subset C^1$.

We derive an a priori estimate of positive solutions of Eq. (3.8). Prior to being able to prove the estimate, we require a technical result.

Lemma 3.12 *Let $u \in H_{\mathcal{B}}^2$ be a solution of Eq. (3.8). Then*

$$u'(x) = \alpha u(x)\,\mathcal{K}[u](x). \tag{3.9}$$

and $u \in C^3(S_L^1)$.

Proof By Sobolev's theorem, we have that $u \in C^{1,1/2}(S_L^1)$. Integrating Eq. (3.8) from $\hat{x}$ (the point at which the flux is zero, whose existence is guaranteed by Remark 3.3), we observe that

$$u'(x) = \alpha u(x)\,\mathcal{K}[u](x). \tag{3.10}$$

Lemma 3.3 implies that whenever $u \in H^2$, we also have that $\mathcal{K}[u] \in H^2$; hence, applying the Banach algebra property of H^2 and Sobolev's theorem, we have that $u\mathcal{K}[u] \in H^2 \subset\subset C^{1,1/2} \subset C^1$. Then by Eq. (3.9), we have that $u' \in C^1(S_L^1)$ and thus $u \in C^2(S_L^1)$. Further from Eq. (3.8), we have that

$$u''(x) = \alpha \, (u(x) \, \mathcal{K}[u](x))'.$$

As $u \in C^2(S_L^1)$, we apply Lemma 3.3 (P1) to find that $\mathcal{K}[u] \in C^2(S_L^1)$; hence, $u\mathcal{K}[u] \in C^2(S_L^1)$. This means that $u'' \in C^1(S_L^1)$ and finally we have that $u \in C^3(S_L^1)$.   □

Lemma 3.13 *Let $u \in C^1(S_L^1)$ be a non-negative solution of Eq. (3.8), subject to the integral constraint*

$$\mathcal{A}[u] = \bar{u},$$

where $\bar{u} > 0$. Then, we have

$$\bar{u}e^{-\alpha\mu L} \leq u(x) \leq \bar{u}e^{\alpha\mu L},$$

where $\mu = C(\bar{u} + L)|\omega|_\infty$, and C from (P6). Further we have that

$$|u'(x)| \leq \alpha\mu\bar{u}e^{\alpha\mu L}.$$

Then,

$$\|u\|_{C^1} \leq (1 + \alpha\mu)\,\bar{u}e^{\alpha\mu L} := \kappa(\alpha, L, \bar{u}, \Omega).$$

Proof Integrating Eq. (3.8) from $\hat{x}$ (given in Remark 3.3) to x, we obtain Eq. (3.9). Thus using Lemma 3.3 (P6), we obtain the following differential inequality:

$$u'(x) \leq \alpha C|\omega|_\infty u(x)\,(\bar{u} + L).$$

Let us denote $\mu = C(\bar{u} + L)|\omega|_\infty$. Next we note that if $u(x)$ has mass $\bar{u}$, then there is $\tilde{x} \in [0, L]$ such that $u(\tilde{x}) = \bar{u}$. Then integrating from x to $\tilde{x}$, we obtain

$$\ln u(\hat{x}) - \ln u(x) \leq \alpha\mu L.$$

Similarly integrating from $\tilde{x}$ to x, we obtain

$$\ln u(x) - \ln u(\tilde{x}) \leq \alpha\mu L.$$

Combining both inequalities, we find

$$-\alpha\mu L \leq \ln u(x) - \ln u(\tilde{x}) \leq \alpha\mu L,$$

which yields the a priori estimate.   □

Next we show that for steady states $u(x)$ of Eq. (3.8), we find that $u(x)$ has maxima or minima whenever $\mathcal{K}[u](x) = 0$.

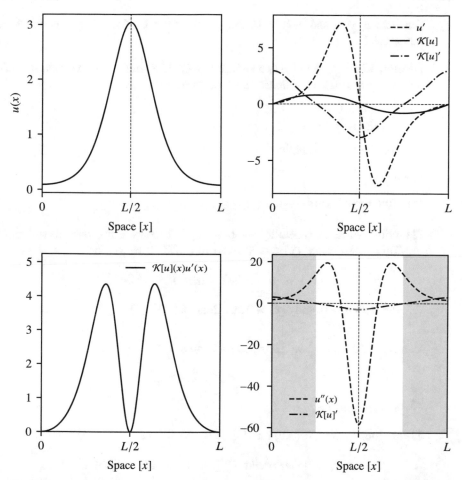

Fig. 3.4 Top Left: a typical non-trivial steady-state solution. Top Right: examples of the non-local term $\mathcal{K}[u]$ and $\mathcal{K}[u]'$ applied to the solution on the left. We observe the properties proven in Lemma 3.14 and Lemma 3.15. We observe that u' and $\mathcal{K}[u]$ have the same positive and negative regions. The dashed black line denotes the locations of the zeros of $\mathcal{K}[u](x)$ and $u'(x)$. Bottom Left: the product of $\mathcal{K}[u](x)$ and $u'(x)$ to show that they have the same sign and the same zeroes. The dashed black line denotes the locations of the zeros of $\mathcal{K}[u](x)$ and $u'(x)$. Bottom Right: a comparison of the second derivative and $\mathcal{K}[u]'$ and u''. If u is convex, it is also non-locally convex, and if u is non-locally concave, then it is concave. The dashed black line denotes the locations of the zeros of $\mathcal{K}[u]$ and u'

Lemma 3.14 *Suppose $u(x)$ is a solution of Eq. (3.8) and assume $u(\hat{x}) > 0$. Then $u'(\hat{x}) = 0$ if and only if $\mathcal{K}[u](\hat{x}) = 0$ (see Fig. 3.4).*

Proof Suppose that at $\hat{x} \in S_L^1$, we have $u'(\hat{x}) = 0$; then Eq. (3.9) implies that

$$0 = \alpha u(\hat{x}) \, \mathcal{K}[u](\hat{x}).$$

But both $\alpha \neq 0$ and $u(\hat{x}) \neq 0$; thus, $\mathcal{K}[u](\hat{x}) = 0$. The other direction follows immediately. □

Lemma 3.15 *Suppose $u(x)$ is a solution of Eq. (3.8) and $u(\hat{x}) > 0$; then it achieves a non-zero maximum (minimum) at $\hat{x}$ if and only if*

  1. *$\mathcal{K}[u](\hat{x}) = 0$, and*
  2. *$(\mathcal{K}[u](\hat{x}))' < (>) 0$.*

See Fig. 3.4 for an example.

Proof

(1) from Lemma 3.14.

(2) Without loss of generality, suppose $u(x)$ achieves a non-zero maximum at $\hat{x}$. This means that $u''(\hat{x}) < 0$. Thus from Eq. (3.8), we get that

$$0 > u''(\hat{x}) = \alpha u(\hat{x}) (\mathcal{K}[u](\hat{x}))' .$$

But both $\alpha > 0$, and $u(\hat{x}) > 0$ and thus $(\mathcal{K}[u])' < 0$. □

Lemma 3.16 *Let $u(x)$ be a positive solution of Eq. (3.8); then*

$$u'(x)\,\mathcal{K}[u](x) \geq 0.$$

For an example, see Fig. 3.4.

Proof Substituting Eq. (3.9) into Eq. (3.8), we obtain

$$u''(x) = \alpha^2 u(x) (\mathcal{K}[u](x))^2 + \alpha u(x) (\mathcal{K}[u])' .$$

Note that the first term on the right-hand side is positive and thus we have that

$$u'' \geq \alpha u(x) (\mathcal{K}[u])' . \tag{3.11}$$

Then using Eq. (3.8) and result (3.11), we obtain that

$$\alpha u'(x)\,\mathcal{K}[u](x) = u''(x) - \alpha u(x)(\mathcal{K}[u](x))' \geq 0.$$

Finally we can relate the local and non-local curvatures for steady-state solutions:

Lemma 3.17 *Let $u(x)$ be a solution of Eq. (3.8); then*

  1. *If $u''(x) \leq 0$, then $(\mathcal{K}[u](x))' \leq 0$,*
  2. *If $(\mathcal{K}[u](x))' \geq 0$, then $u''(x) \geq 0$.*

Proof Proof is by Eq. (3.11). □

3.6 Summary

In this section, we studied basic mathematical properties of the non-local operator $\mathcal{K}[u]$, such as continuity, regularity, growth estimates, and spectral properties (linear non-local operator). These results are relevant for subsequent bifurcation analysis.

We saw that the linear non-local operator shares many properties with the first derivative operator. It is skew-adjoint, it has the complex exponentials as eigenfunctions with purely complex eigenvalues, and it maps sine basis functions to cosine basis functions and cosine basis functions to negative sine. The range of $\mathcal{K}[u]$ is contained in the subspace of zero average functions, and the nullspaces are all the constant functions. There are however some differences as well. Most notably $\mathcal{K}[u]$ is a compact operator; thus, its eigenvalues accumulate at zero (see Example 3.1), while the eigenvalues of $(\cdot)'$ diverge.

Similarly, the non-local curvature $\mathcal{K}[u]'$ shares many properties with a second derivative $(\cdot)''$. Both are self-adjoint, and both share the same eigenfunctions (provided we consider $(\cdot)''$ with periodic boundary conditions as well). Differences are that $\mathcal{K}[u]'$ is always a bounded operator, with bounded eigenvalues. The eigenvalues of the second derivative, $-\frac{2\pi n}{L}$, are modified by a specific factor of $2M_n(\omega)$, with

$$M_n(\omega) = \int_0^1 \sin\left(\frac{2n\pi r}{L}\right) \omega(r)\,dr,$$

which are the Fourier sine coefficients of a Fourier sine expansion of $\omega(r)$.

In this section, we have seen that the non-local operator $\mathcal{K}[u]$ can be viewed as a generalization of the local derivative (in the sense that as $R \to 0$, $\mathcal{K}_R[u] \to c_1 u'$).

In view of this analogy, the estimate $(|\mathcal{K}[u]|_p \leq C(|u|_p + L))$ obtained in Lemma 3.3 (P3) can be viewed as a type of reverse Poincaré inequality. Intuitively this estimate "earns" us an order of differentiability. Similar results for other non-local operators have been obtained in [100, 108, 109].

Chapter 4
Local Bifurcation

The success of the Armstrong–Painter–Sherratt adhesion model (2.14) is that it can replicate the complicated patterns observed in cell sorting experiments [10]. In mathematical terms, these patterns are steady states of Eq. (2.14). Thus, understanding the conditions under which these steady states form and become stable is important. Furthermore, understanding the set of steady states is one of the first steps toward understanding the equation's global attractor.

Up to this point, the steady states of Eq. (2.14) have only been studied numerically and using linear stability analysis [10]. Closely related to Eq. (2.14) are the local and non-local chemotaxis equations [100, 101, 143]. For both the local and non-local chemotaxis equations, a global bifurcation analysis, to understand their steady states, was carried out [184, 192]. Inspired by their results, we present here an exploration of the set of non-homogeneous steady-state solutions of Eq. (2.14) in one dimension, i.e., Eq. (3.1).

Central to our analysis is the abstract bifurcation theory of Crandall and Rabinowitz [46, 156], in particular the local bifurcation Theorem 2.3 (see [46]) and the global bifurcation Theorem 2.7 (see [156]). Preceding the formulation of these general theorems, Crandall et al. studied the set of solutions of nonlinear Sturm–Liouville problems [45, 155]. For linear Sturm–Liouville eigenvalue problems, it is well known that the eigenfunctions can be classified by their number of zeros [43]. Crandall et al. [46] showed that under rather weak assumptions the same classification holds for nonlinear eigenvalue problems. In fact, each global solution branch inherits the number of zeros of the eigenfunction spanning the nullspace at the bifurcation point. Furthermore, they showed that each global solution branch is unbounded and that branches do not meet (since it is impossible for solutions of Sturm–Liouville problems to have degenerate zeros). In [156], Rabinowitz formulates a general global bifurcation theorem, providing two alternatives for the global structure of the solution branches. They either are bounded, connecting two bifurcation points, or are unbounded (see Theorem 1.3 in [156] and Theorem 2.7 in Chap. 2). This is now known as the *Rabinowitz-alternative*. An extension of the global bifurcation theorem to study so-called unilateral branches (sub-branches in only the positive or negative direction of the eigenfunction at the bifurcation point)

© The Author(s), under exclusive license to Springer Nature Switzerland AG 2021
A. Buttenschön, T. Hillen, *Non-Local Cell Adhesion Models*, CMS/CAIMS Books in
Mathematics 1, https://doi.org/10.1007/978-3-030-67111-2_4

was originally reported in [156]. The original proof contained holes that were filled in by Dancer [48], López-Gómez [120, 121], and Shi and Wang [167]. Since the original formulation of these bifurcation theorems, similar theorems, which apply in more general settings, have been developed. Here, we will use bifurcation theorems that are applicable to Fredholm operators [120, 121, 167] (see Chap. 5).

While both the local and non-local chemotaxis models in [184] and [192] were formulated with no-flux boundary conditions, the formulation of no-flux boundary conditions for Eq. (3.1) is more challenging due to the complicated non-local structure of $\mathcal{K}[u]$. Hence, we first study and analyze the formation of non-homogeneous solutions in isolation of boundary considerations, and we formulate Eq. (3.1) on a circle (equivalent to periodic boundary conditions). A more detailed discussion on the challenges of construction of no-flux boundary conditions can be found in Part III.

At the same time however, formulating our model on a circle gives rise to some additional challenges. The most critical being that eigenvalues of the Laplacian on S_L^1 need not be simple. This is a challenge, because bifurcations require eigenvalues of odd multiplicity (the theorems in [121, 167] require simple eigenvalues). This challenge was previously observed by Matano, who studied nonlinear reaction diffusion equations on the circle [126]. The solution of Matano [126] was to impose symmetry requirements on the nonlinear term, so that the equation would be $\mathbf{O}(2)$ equivariant, i.e., invariant under translations and reflections (see also Sect. 4.2). This will also be our approach here.

Around the same time as Matano [126], Healey [93] extended the Rabinowitz alternative to so-called G-reduced problems (G being a symmetry group), which is a nonlinear bifurcation problem formulated on a Banach space whose elements are fixed points of the symmetries defined by G [93], thus showing that certain symmetries persist along global bifurcation branches (similar results are often referred to as the equivariant branching lemma [82]). Subsequently, these ideas were used in a series of papers, which studied the condition under which solutions of nonlinear elliptic equations are classifiable by their number of zeros. The key to these results was to impose sufficient conditions on the nonlinear terms such that the resulting equation was equivariant under actions of $\mathbf{O}(2) \times \mathbb{Z}_2$ ($\mathbb{Z}_2$ describes the action of the negative identity, i.e., a reflection through the x-axis) [94, 95]. Intuitively, this symmetry requirement ensured that the zeros of solutions were "frozen" (i.e., fixed location), and thus the number of zeros is preserved along the global bifurcation branch. Furthermore, this result easily shows that global solution branches do not meet. Much more recently, Buono et al. used $\mathbf{O}(2)$ equivariance to compare the accessible bifurcations in a non-local hyperbolic model of swarming and the equation's formal parabolic limit [28]. In a similar spirit, we will show that the steady-state equation (4.1) of Eq. (3.1) is indeed equivariant under actions of $\mathbf{O}(2)$. Using the properties of the non-local term $\mathcal{K}[u]$, we will then show that this leads to "frozen" maxima and minima (equivalently frozen zeros of the derivative u'). Since we consider Eq. (4.1) on a periodic domain, we can prescribe the location of one maxima (or minima) without restricting possible solutions. This ensures that only simple eigenvalues occur in our subsequent analysis. In Chap. 5, we prove a global bifurcation result for the steady-state solutions of the non-local cell-cell adhesion equation.

4.1 The Abstract Bifurcation Problem

The steady states of Eq. (3.1) are solutions of the following non-local equation

$$u''(x) = \alpha \Big(u(x)\, \mathcal{K}[u(x)](x) \Big)' \quad \text{in } S_L^1. \tag{4.1a}$$

Some typical steady-state solutions of this equation are shown in Fig. 3.1. Due to the mass conserving property of Eq. (3.1), we find families of steady states that are parameterized by their average

$$\bar{u} = \mathcal{A}[u] = \frac{1}{L} \int_0^L u(x)\, dx. \tag{4.1b}$$

We understand this equation as constraint, which ensures that bifurcations arise in a given family of solutions which have the same mass (note that the mass is $L\bar{u}$). We define the following function space:

$$H_{\mathcal{B}}^2(S_L^1) := \big\{ u \in H^2(S_L^1) : \mathcal{B}[u, u'] = 0 \big\},$$

where the boundary operator $\mathcal{B}[\cdot, \cdot]$ was defined in Eq. (2.11) for periodic boundary conditions. For given $\bar{u} > 0$, we define the following operator:

$$\mathcal{F} : \mathbb{R} \times H_{\mathcal{B}}^2(S_L^1) \mapsto L^2(S_L^1) \times \mathbb{R}, \tag{4.2a}$$

$$\mathcal{F}[\alpha, u] = \begin{bmatrix} (-u' + \alpha u\, \mathcal{K}[u])' \\ \mathcal{A}[u] - \bar{u} \end{bmatrix}. \tag{4.2b}$$

We will need the Fréchet derivative of $\mathcal{F}$.

Lemma 4.1 *The Fréchet derivative* $\mathcal{D}_u \mathcal{F} : \mathbb{R} \times H_{\mathcal{B}}^2 \mapsto \mathfrak{L}(H_{\mathcal{B}}^2, L^2 \times \mathbb{R})$ *of the operator* $\mathcal{F}$ *is given by*

$$\mathcal{D}_u \mathcal{F}(\alpha, u)[w] = \begin{pmatrix} [-w' + \alpha (u\, \mathcal{K}_h[w] + w\, \mathcal{K}[u])]' \\ \mathcal{A}[w] \end{pmatrix}. \tag{4.3}$$

where

$$\mathcal{K}_h[w] = \int_{-1}^1 h'(u(x + r))w(x + r)\Omega(r)\, dr. \tag{4.4}$$

Proof Let $u, v, w \in H_{\mathcal{B}}^2$ and we compute

$$\mathcal{F}[\alpha, u + w] - \mathcal{F}[\alpha, u] - \mathcal{D}_u \mathcal{F}(\alpha, u)[w] \tag{4.5}$$

The second component is

$$\mathcal{A}[u + w] - \bar{u} - (\mathcal{A}[u] - \bar{u}) - \mathcal{A}[w] = 0.$$

In the first component of (4.5), the local derivative terms cancel directly:

$$(-(u + w)' - (-u') - (-w'))' = 0.$$

It remains to check the non-local terms in the first component of (4.5):

$$\left(\mathcal{F}[\alpha, u + w] - \mathcal{F}[\alpha, u] - \mathcal{D}_u \mathcal{F}(\alpha, u)[w] \right)_1$$

$$= \alpha \left(u \, \mathcal{K}[u + w] + w \, \mathcal{K}[u + w] - u \, \mathcal{K}[u] - u \, \mathcal{K}_h[w] + w \, \mathcal{K}[u] \right)'$$

$$= \alpha \left[\left(u \underbrace{\int_{-1}^{1} (h(u + w) - h(w) - h'(u)w) \, \Omega(r) \, dr}_{\text{I}} \right)' \right.$$

$$\left. + \left(w \underbrace{\int_{-1}^{1} (h(u + w) - h(u)) \, \Omega(r) \, dr}_{\text{II}} \right)' \right].$$

We require an L^2 estimate of the previous terms. For this, we consider the two terms separately (we denote them by I and II, respectively).

$$|(\text{I})'|_2 \leq \|\text{I}\|_{H^1} \leq \|u\|_{H^1} \|w\|_{H^1} \left\| \frac{h(u + w) - h(u) - h'(u)w}{w} \right\|_{H^1},$$

where we use the Banach algebra property of H^1. Now, by assumption H1 $h(\cdot) \in C^2$; hence, we have that

$$\lim_{\|w\| \to 0} \left\| \frac{h(u + w) - h(u) - h'(u)w}{w} \right\|_{H^1} = 0.$$

For the second term, we proceed similarly

$$|(\text{II})'|_2 \leq \|\text{II}\|_{H^1} \leq \|w\|_{H^1} \|h(u + w) - h(u)\|_{H^1},$$

which goes to 0 as $\|w\| \to 0$, since $h(u)$ is continuous. Together we find

$$\frac{|\mathcal{F}[\alpha, u+w] - \mathcal{F}[\alpha, u] - \mathcal{D}_u \mathcal{F}(\alpha, u)[w]|_2}{\|w\|_{H^2}}$$

$$\leq \frac{\|w\|_{H^1}}{\|w\|_{H^2}} \alpha \|u\|_{H^1} \left(\left\| \frac{h(u+w) - h(u) - h'(u+w)}{w} \right\|_{H^1} + \|h(u+w) - h(w)\|_{H^1} \right)$$

$$\to 0 \quad \text{as} \quad \|w\|_{H^2} \to 0.$$

Hence, $\mathcal{D}_u \mathcal{F}(\alpha, u)$ is the Fréchet derivative of $\mathcal{F}$. $\qquad \square$

We then prove a series of properties of $\mathcal{F}$ that allow us to apply the bifurcation theorems from Chap. 2.

Lemma 4.2 *For each $\mathbb{R} \ni \bar{u} > 0$, the operator $\mathcal{F}$ as defined in Eq. (4.2) has the following properties:*

1. $\mathcal{F}[\alpha, \bar{u}] = 0$ *for all $\alpha \in \mathbb{R}$.*
2. *The first component of $\mathcal{F}$ maps into $L_0^2(S_L^1)$ (defined in Eq. (2.13)).*
3. $\mathcal{D}_u \mathcal{F}(\alpha, u)$ *is Fredholm with index 0, for each $\alpha \in \mathbb{R}$.*
4. $\mathcal{F}[\alpha, u]$ *is C^1 smooth in u.*
5. $\mathcal{D}_{\alpha u} \mathcal{F}(\alpha, u)$ *exists and is continuous in u.*

Proof

1. We note that $\mathcal{K}[\bar{u}] = 0$. Hence, the conclusion follows.
2. This is easily seen by integrating the equation and using the periodic boundary conditions $\mathcal{B}$.
3. The Fréchet derivative of $\mathcal{F}$ with respect to u was found in Lemma 4.1 to be

$$\mathcal{D}_u \mathcal{F}(\alpha, u)[w] = \begin{pmatrix} (-w' + \alpha (u \mathcal{K}_h[w] + w \mathcal{K}[u]))' \\ \mathcal{A}[w] \end{pmatrix}.$$

We split this operator as follows:

$$\mathcal{D}_u \mathcal{F}(\alpha, u)[w] = \mathcal{T}_1(\alpha, u)[w] + \mathcal{T}_2(\alpha, u)[w],$$

where

$$\mathcal{T}_1(\alpha, u)[w] = \begin{pmatrix} -w'' \\ 0 \end{pmatrix},$$

and

$$\mathcal{T}_2(\alpha, u)[w] = \begin{pmatrix} \alpha (u \mathcal{K}_h[w] + w \mathcal{K}[u])' \\ \mathcal{A}[w] \end{pmatrix}.$$

$\mathcal{T}_1(\alpha, u)$ is Fredholm with index 0, by Lemma 2.5.
The operator $\mathcal{T}_2(\alpha, u)$ is compact. First, consider the first component of $\mathcal{T}_2$. Now $\mathcal{D} : H^1 \mapsto L^2, w \mapsto w'$ is continuous. Then let $\mathcal{T}(u)[w] = u \mathcal{K}_h[w] + w \mathcal{K}[u]$,

where $u, w \in H_{\mathcal{B}}^2$. Since $h'(u) \in C^1$, we have that $\mathcal{K}_h[w] \in H_{\mathcal{B}}^2$. Since $H_{\mathcal{B}}^2$ is a Banach algebra, we have that $u\,\mathcal{K}_h[w] \in H_{\mathcal{B}}^2$ and $w\,\mathcal{K}[u] \in H_{\mathcal{B}}^2$. Since $H^2 \subset\subset C^1$, we can conclude that the first component is given by the composition $\mathcal{D} \circ \mathcal{T}(u)$ and hence is compact.

For the second component, we let $(w_n) \subset H^2$ be bounded; then $\mathcal{A}[w_n] = \frac{1}{L} \int_0^L w_n(x)\,dx$, but as $(w_n) \subset L^\infty$, we have that $\mathcal{A}[w_n]$ is a bounded sequence in $\mathbb{R}$ and thus has a convergent subsequence, making $\mathcal{A}$ compact.

Finally, we recall the well-known results that the compact perturbation of a Fredholm operator is Fredholm with the same index (see Theorem 2.2).

Hence, $\mathcal{D}_u\,\mathcal{F}(\alpha, u)$ is Fredholm with index 0.

4. For this, we have to check that the mapping

$$u \mapsto \mathcal{D}_u\,\mathcal{F}(\alpha, u)$$

is continuous. The u dependence in the linearization $\mathcal{D}_u\,\mathcal{F}(\alpha, u)$ arises in the integral terms. For each $w \in H^2$, the map

$$u \mapsto \mathcal{D}_u\,\mathcal{F}(\alpha, u)[w] = \begin{pmatrix} -w'' + \alpha(u\,\mathcal{K}_h[w] + w\,\mathcal{K}[u])' \\ \mathcal{A}[w] \end{pmatrix}$$

is continuous in u, since $\mathcal{K}_h[w]$ is bounded and $\mathcal{K}[u]$ is continuous by Lemma 3.3.

5. We simply compute to find

$$\mathcal{D}_{\alpha u}\,\mathcal{F}(\alpha, u) = \begin{pmatrix} (u\,\mathcal{K}_h[w] + w\,\mathcal{K}[u])' \\ 0 \end{pmatrix}.$$

Its continuity follows from item (4).   □

Remark 4.1 Note that Lemma 4.2 implies that $\mathcal{F}$ satisfies properties (F1), (F2), and (F3) from Sect. 2.3.

4.2 Symmetries and Equivariant Flows

For the following bifurcation analysis, we will require a detailed understanding of symmetries of our solutions. We use group theory to describe these symmetries. We obtain the following result.

Lemma 4.3 *The operator $\mathcal{F}$ defined in Eq. (4.2b) is equivariant under the actions of $\mathbf{O}(2)$, i.e.,*

$$\mathcal{F}[\alpha, \gamma u] = \gamma\,\mathcal{F}[\alpha, u], \quad \forall \gamma \in \mathbf{O}(2).$$

Remark 4.2 The group $\mathbf{O}(2)$ is generated by $\mathbf{SO}(2)$ and a reflection. In more detail, $\mathbf{SO}(2)$ can be represented in $\mathbb{R}^2$ by rotations

$$\sigma_\theta = \begin{pmatrix} \cos\theta & -\sin\theta \\ \sin\theta & \cos\theta \end{pmatrix}, \quad \theta \in [0, 2\pi)$$

and a reflection given by

$$\rho = \begin{pmatrix} 1 & 0 \\ 0 & -1 \end{pmatrix}.$$

It is easy to see that this group is compact and hence proper. Here we represent the group by its action on functions on S_L^1 as

$$\begin{aligned} \sigma_a u(x) &= u(x - a) \quad a \in [0, L] \quad \text{translations,} \\ \rho u(x) &= u(L - x) \quad \text{reflection.} \end{aligned} \tag{4.6}$$

In the following, we will denote reflection about $a \in S_L^1$ by

$$\rho_a u(x) = u(2a - x),$$

Since $\rho_a = \sigma_{L-2a}\rho$, this operation is in $\mathbf{O}(2)$. See examples in Fig. 4.1.

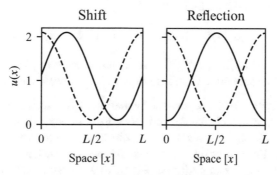

Fig. 4.1 Examples of the actions of σ_a (left) and ρ_a (right). Here $a = {}^L/_4$. In both subplots, the original function is shown in dashed black, while the shifted and, respectively, reflected functions are solid

For the proof of this lemma, we require the following lemma describing how the non-local operators behave under actions of $\mathbf{O}(2)$.

Lemma 4.4 *Let the non-local operator $\mathcal{K}[u]$ be defined as in Definition 3.1; then*

$$\mathcal{K}[\sigma_\theta u] = \sigma_\theta \, \mathcal{K}[u], \qquad \mathcal{K}[\rho_a u] = -\rho_a \, \mathcal{K}[u],$$

and for the non-local curvature operator, we have that

$$(\mathcal{K}[\sigma_\theta u])' = \sigma_\theta (\mathcal{K}[u])', \qquad (\mathcal{K}[\rho_a u])' = \rho_a (\mathcal{K}[u])'.$$

Proof The results for σ_θ are trivial. For ρ_a, we first deal with $\mathcal{K}[u](x)$. Then

$$\mathcal{K}[\rho_a u](x) = \int_0^1 [\, h(u(2a - x - r)) - h(u(2a - x + r))\,]\, \omega(r)\, dr$$
$$= -\mathcal{K}[u](2a - x) = -\rho_a (\mathcal{K}[u]).$$

Second, we show the same for $\mathcal{K}[u]'$ using a simple change of variables

$$\mathcal{K}[\rho_a u]' = \left(\int_{-1}^1 -h(u(2a - x - r))\Omega(r)\, dr \right)'$$
$$= \left(\int_{-1}^1 h(u(2a - x + y))\Omega(y)\, dy \right)' = \mathcal{K}[u]'(2a - x) = \rho_a \mathcal{K}[u]'.$$

Now we can complete the proof of Lemma 4.3.

Proof of Lemma 4.3 The elements of **SO**(2) are given by the translations. As

$$\frac{\partial}{\partial x} u(x - a) = \frac{\partial}{\partial(x - a)} u(x - a),$$

it is trivial to see that $\mathcal{F}$ is equivariant under actions of elements in **SO**(2). Note that to obtain all elements in **O**(2), we only need the reflection through $a = {}^L/_2$ as defined in Eq. (4.6); we find that

$$(\rho v(x))' = (v(2a - x))' = -v'(2a - x) = -(\rho v')(x),$$

and hence $(\rho v(x))'' = (\rho v'')(x)$. For the non-local term, we apply Lemma 4.4 with $a = {}^L/_2$. Then substitute these into Eq. (4.2b), and we obtain the result. $\qquad\square$

The linearization of the steady-state equation (4.1a) is given by $\mathcal{D}_u \mathcal{F}(\alpha, u)[w] = 0$. We saw in Lemma 3.6 that the non-local operator $\mathcal{K}$ maps $\sin(\frac{n\pi x}{L})$ to $\cos(\frac{n\pi x}{L})$ and vice versa. Hence, as we will show below, the eigenfunctions of the linearization $\mathcal{D}_u \mathcal{F}$ are sine and cosine functions. As we are working on the circle of length L, a sine function can easily be shifted to a cosine function with an appropriate phase shift. For the bifurcation analysis, we need to remove this symmetry, which we achieve by stipulating the reflection symmetry $u(x) = u(L - x)$. This makes the corresponding eigenspaces one-dimensional, and we can apply the abstract bifurcation theory outlined before. Once these bifurcation branches are identified, we can shift the solutions around the circle to obtain other solutions that are not reflection symmetric through the domain's center.

For this reason, we define new function spaces

$$H_P^2 := \left\{ u \in H_{\mathcal{B}}^2(S_L^1) : \; u(x) = u(L - x) \right\} \tag{4.7}$$

and L_P^2 accordingly. It is then easy to see that the operator $\mathcal{F}$ now maps $\mathbb{R} \times H_P^2 \mapsto L_P^2 \times \mathbb{R}$, since if $u \in H_P^2$ we find that

$$\mathcal{F}[\alpha, u] = \mathcal{F}[\alpha, \rho u] = \rho \mathcal{F}[\alpha, u].$$

4.3 Singular Points of $\mathcal{F}$

Due to Lemma 2.3, we have that possible bifurcation points are those that are singular points of the linearization of $\mathcal{F}$ evaluated at the trivial solution $(\alpha, \bar{u})$. The main bifurcation result of Crandall and Rabinowitz Theorem 2.3 is formulated for a trivial steady state at 0. Hence, we need to shift our solutions by $-\bar{u}$ to obtain a trivial steady state at 0.

$$v(x) := u(x) - \bar{u}. \tag{4.8}$$

Under this change of variable, the operator $\mathcal{F}$ on H_P^2 becomes

$$\mathcal{F}[\alpha, v] = \begin{bmatrix} (-v' + \alpha(v + \bar{u}) \, \mathcal{K}[v])' \\ \mathcal{A}[v] \end{bmatrix}, \tag{4.9}$$

and its linearization becomes

$$\mathcal{D}_v \, \mathcal{F}(\alpha, v)[w] = \begin{pmatrix} (-w' + \alpha \, ((v + \bar{u}) \, \mathcal{K}_h[w] + w \, \mathcal{K}[v]))' \\ \mathcal{A}[w] \end{pmatrix}.$$

Note that we treat the linearizations as an operator family indexed by $\alpha \in \mathbb{R}$. In the next two lemmas, we characterize the values of α which lie in the singular set $\mathrm{Sing}\,(\mathcal{D}_v \, \mathcal{F}(\alpha, 0))$.

Lemma 4.5 *Let the operator $\mathcal{F}$ on H_P^2 be as defined in (4.9), and let $\alpha > 0$. Its Fréchet derivative has been shown to exist in Lemma 4.2. Assume that $M_n(\omega) > 0$ (i.e., the Fourier sine coefficients of ω as defined in Eq. (3.4)), and define*

$$\alpha_n := \frac{n\pi}{\bar{u} L M_n(\omega) h'(\bar{u})} \quad \text{for } n \in \mathbb{N}. \tag{4.10}$$

Then, we have that

$$\dim \mathrm{N}[\mathcal{D}_v \, \mathcal{F}(\alpha_n, 0)] = 1.$$

Thus, the singular points of the linearization are given by the α_n:

$$\mathrm{Sing}\,(\mathcal{D}_v \, \mathcal{F}(\alpha, 0)) = \{\, \alpha_n \, : \, n \in \mathbb{N} \setminus \{0\} \,\}.$$

Remark 4.3 Note that $\alpha = 0$ is not an eigenvalue of $\mathcal{D}_v \mathcal{F}(\alpha, 0)$, since in this case, the only solution of Eq. (4.11) is the zero solution.

Remark 4.4 Note that since we assume that $L \geq 2$, we have that $M_1(\omega) > 0$; since then, we have that $\sin\left(\frac{2\pi x}{L}\right) > 0$ on $(0, 1)$ and $\omega(r) \geq 0$ by assumption. Thus, there is always one such bifurcation point.

Proof The nullspace of $\mathcal{D}_v \mathcal{F}(\alpha, 0)$ is given by the solution of the following equation:

$$\begin{cases} -w'' + \alpha \bar{u} h'(\bar{u}) \left(\int_{-1}^{1} w(x + r)\Omega(r)\,dr \right)' = 0 \text{ in } [0, L] \\ \mathcal{B}[w, w'] = 0, \quad \mathcal{A}[w] = 0. \end{cases} \tag{4.11}$$

We solve this system using an eigenfunction ansatz.

$$w(x) = a_0 + \sum_{n=1}^{\infty} a_n \cos\left(\frac{2n\pi x}{L}\right) + \sum_{n=1}^{\infty} b_n \sin\left(\frac{2n\pi x}{L}\right),$$

then because $\mathcal{A}[w] = 0$, we have that $a_0 = 0$, and as $w \in H_P^2$ has reflection symmetry, we have that $b_n = 0, \forall n \in \mathbb{N}$. Hence

$$w(x) = \sum_{n=1}^{\infty} a_n \cos\left(\frac{2n\pi x}{L}\right),$$

which is then substituted into Eq. (4.11) to obtain

$$\sum_{n=1}^{\infty} a_n \frac{2\pi n}{L} \cos\left(\frac{2n\pi x}{L}\right) \left\{ 2\alpha \bar{u} h'(\bar{u}) \int_{0}^{1} \sin\left(\frac{2n\pi x}{L}\right) \omega(r)\,dr - \frac{2n\pi}{L} \right\} = 0.$$

To find a non-trivial solution, we require that

$$\left\{ \alpha \bar{u} h'(\bar{u}) \int_{0}^{1} \sin\left(\frac{2n\pi x}{L}\right) \omega(r)\,dr - \frac{n\pi}{L} \right\} = 0.$$

Hence, $\mathcal{D}_v \mathcal{F}(\alpha, 0)$ is not an isomorphism whenever α equals one of the following:

$$\alpha_n = \frac{n\pi}{L\bar{u}M_n(\omega)h'(\bar{u})},$$

where $M_n(\omega)$ is defined in Eq. (3.4). The corresponding eigenfunctions are

$$e_n(x) = \cos\left(\frac{2\pi n x}{L}\right), \quad n = 1, 2, \ldots.$$

Remark 4.5 Note that the values α_n found in Lemma 4.5 are exactly the values at which an eigenvalue λ of the linear operator

$$v'' - \alpha \bar{u} \, (\mathcal{K}[u])' = \lambda v, \tag{4.12}$$

crosses through 0.

Lemma 4.6 *Let α_n be a generalized eigenvalue of $\mathcal{D}_v \, \mathcal{F}(\alpha, 0)$ as found in Lemma 4.5. Then, we have that*

$$\mathcal{D}_{\alpha v} \, \mathcal{F}(\alpha_n, 0)[e_n] \notin \mathrm{R}[\mathcal{D}_v \, \mathcal{F}(\alpha_n, 0)],$$

where $e_n(x)$ is the eigenfunction corresponding to α_n from (4.10).

Proof We proceed by contradiction. We assume that

$$\mathcal{D}_{\alpha v} \, \mathcal{F}(\alpha_n, 0)[e_n] \in \mathrm{R}[\mathcal{D}_v \, \mathcal{F}(\alpha_n, 0)]. \tag{4.13}$$

That means that there is $z \in H_{\mathcal{B}}^2$ such that

$$\mathcal{D}_v \, \mathcal{F}(\alpha_n, 0)[z] = \mathcal{D}_{\alpha v} \, \mathcal{F}(\alpha_n, 0)[e_n] = \begin{bmatrix} \bar{u} \mathcal{K}_h[e_n]' \\ 0 \end{bmatrix}. \tag{4.14}$$

Then Eq. (4.14) is equivalent to

$$\begin{cases} -z'' + \alpha_n \bar{u} h'(\bar{u}) \left(\int_{-1}^{1} w(x + r)\Omega(r) \, dr \right)' = \bar{u} \mathcal{K}_h[e_n]' \text{ in } [0, L] \\ \mathcal{B}[z, z'] = 0, \quad \mathcal{A}[z] = 0. \end{cases} \tag{4.15}$$

We note that (refer to Lemma 3.8)

$$\mathcal{K}_h[e_n]' = -h'(\bar{u}) \left(\frac{4\pi n}{L} \right) \cos \left(\frac{2\pi n x}{L} \right) M_n(\omega).$$

Like in the previous lemma, we once again use the ansatz

$$z(x) = \sum_{j=1}^{\infty} t_j \cos \left(\frac{2\pi n x}{L} \right).$$

Upon substitution into Eq. (4.15), we obtain

$$\sum_{j=1}^{\infty} t_j \left(\frac{4\pi j}{L} \right) \cos \left(\frac{2\pi j x}{L} \right) \left\{ M_j(\omega) \alpha_n \bar{u} h'(\bar{u}) - \frac{\pi j}{L} \right\}$$

$$= -\bar{u} h'(\bar{u}) \left(\frac{4\pi n}{L} \right) \cos \left(\frac{2\pi n x}{L} \right) M_n(\omega).$$

Then this equation has a solution if and only if $t_j = 0, \forall j \neq n$ and

$$t_n \left\{ \alpha_n M_n(\omega) \bar{u} h'(\bar{u}) - \frac{\pi n}{L} \right\} = -\bar{u} h'(\bar{u}) M_n(\omega)$$

is satisfied. But $\alpha_n = \frac{\pi n}{\bar{u}h'(\bar{u})M_n(\omega)L}$; hence, the term on the left-hand side is zero, and we have found a contradiction.   □

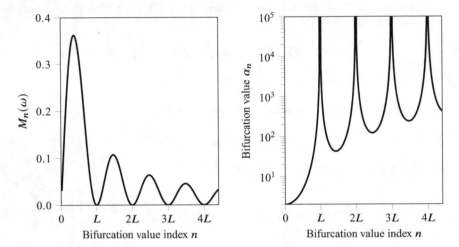

Fig. 4.2 Left: shows the function $M_n(\omega)$ defined in Eq. (3.4) for uniform ω (see O1). When $L = n$, the term $M_n(\omega) = 0$. Right: the bifurcation values α_n as given in Eq. (4.10). The bifurcation values blow up whenever $L = n$. Note that even if $L \neq n$, we still observe very large bifurcation values for $n \sim kL$ ($k \in \mathbb{N}$)

Example 4.1 Suppose that $h(u) = u$ and that ω is uniform. Using the precise form for $M_n(\omega)$ from Example 3.1, the bifurcation values are

$$\alpha_n = \frac{2}{\bar{u}} \left(\frac{L}{n\pi} \sin\left(\frac{n\pi}{L}\right) \right)^{-2}.$$

This means that α_n is a function of $L, \bar{u}, n$. We investigate now the behavior of α_n as a function of L and $\bar{u}$. The domain of L is $[2, \infty)$. When $L = 2$,

$$\alpha_n = \begin{cases} \dfrac{n^2 \pi^2}{2\bar{u}} & \text{if } n \text{ even} \\ 0 & \text{else} \end{cases},$$

and when $L \to \infty$, we have that α_n decreases toward $2/\bar{u}$. When $\bar{u} \to \infty$, we find that $\alpha_n \to 0$. That means that increasing the domain size (equivalently decreasing the sensing radius) or increasing the total mass in the system decreases the threshold adhesion strength required for aggregation. Finally, we note that the map $n \mapsto \alpha_n$ is not monotone, because $M_n(\omega)$ approaches zero when n is close to a multiple of L (see the left side of Fig. 4.2).

4.4 Local Bifurcation Result

Based on the previous lemmas, we can now formulate the local bifurcation result.

Theorem 4.1 *Let $\mathcal{F}$ be given on H_P^2 as in (4.2). Then by Lemmas 4.2, 4.5, and 4.6, all requirements of Theorem 2.3 are satisfied. Then there are continuous functions*

$$(\alpha_k(s), u_k(s)) : (-\delta_k, \delta_k) \to \mathbb{R} \times H_P^2, \qquad (4.16)$$

with $\alpha_k(0) = \alpha_k$ such that

$$u_k(x, s) = \bar{u} + s\alpha_k \cos\left(\frac{2\pi k x}{L}\right) + o(s), \qquad (4.17)$$

where $(\alpha_k(s), u_k(s))$ is a solution of the steady state equation (4.1) and all non-trivial solutions near the bifurcation point $(\alpha_k, \bar{u})$ lie on the curve $\Gamma_k = (\alpha_k(s), u_k(s))$.

Proof To be able to apply Theorem 2.3, we set $U = H_P^2$ and $V = L_{0,P}^2 \times \mathbb{R}$ ($L_{0,P}^2$ is the sub-space of functions in L^2 with the symmetry as in Eq. (4.7) and average zero as in Eq. (2.13)) and $W = H_P^2$. Then, the operator defined in Eq. (4.9) satisfies all the properties required by Theorem 2.3. These properties are proved by Lemma 4.2, Lemma 4.5, and Lemma 4.6. Finally, we revert the change of variables given in Eq. (4.8). To compare the true and the approximate solution see Fig. 4.3.   □

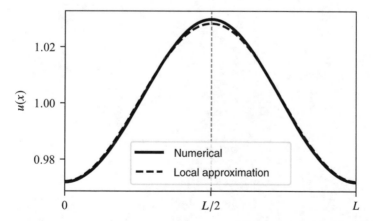

Fig. 4.3 Comparison of numerical solution to the solution approximation given in Eq. (4.17) near the first bifurcation point, i.e., $(\alpha_1, \bar{u})$. The correct value of s for Eq. (4.17) was estimated using the asymptotic expansion introduced in Section 5.3. Here $L = 5$

4.5 Summary

In this chapter, we identified the points at which the trivial steady state $\bar{u}$ of the non-local cell-cell adhesion model (4.1) bifurcates to non-homogeneous solutions. To apply the abstract bifurcation results outlined in Chap. 2, we cast Eq. (4.1) as an abstract operator equation $\mathcal{F}[\alpha, u] = 0$ (see Eq. (4.2)), whose linearization is shown to be a Fredholm operator with index zero. To ensure that the eigenspaces of the linearization $\mathcal{D}_u \mathcal{F}(\alpha, u)$ at possible bifurcation points are one-dimensional, we imposed that solutions be reflection symmetric through the domain's center. Since the domain is periodic, this does not reduce the size of $\mathcal{F}^{-1}(0)$. (Indeed, it's equivalent to the well-known decomposition of any periodic function $u \in H^1$ into $u = \bar{u} + v$ where $v \in H_0^1$.) Interestingly, the existence of bifurcation points depends on the integration kernel of the non-local operator $\mathcal{K}[u]$, since bifurcation points only exist when the quantity $M_n(\omega)$ is positive. Using the previously established properties of the non-local operator $\mathcal{K}$, we are able to apply the abstract local bifurcation theorem of Crandall et al. (Theorem 2.3).

Chapter 5
Global Bifurcation

For each $\bar{u} > 0$, we found local bifurcations at $(\alpha_n, \bar{u})$ with non-trivial eigenfunctions e_n of $\mathcal{D}_u \mathcal{F}(\alpha_n, \bar{u})$ in H_P^2 be given by

$$\alpha_n = \frac{n\pi}{\bar{u}h'(\bar{u})LM_n(\omega)}, \qquad e_n(x) = \cos\left(\frac{2\pi n x}{L}\right), \tag{5.1}$$

where $M_n(\omega)$ are the Fourier sine coefficients of ω (see 3.4). For many examples of PDEs [82, 93–95], the symmetries of the unstable modes e_n are conserved along the bifurcating solution branch. We will show that this is the case here as well. For this, we define the so-called isotropy subgroup associated with e_n. The isotropy subgroup contains all the $\mathbf{O}(2)$-group actions that leave e_n invariant.

$$\Sigma_n := \{ \gamma \in \mathbf{O}(2) : \gamma e_n = e_n \}.$$

It is easy to see that for the eigenfunction e_n, the isotropy subgroup is given by

$$\Sigma_n := \left\{ \sigma_{\frac{mL}{n}}, \; \rho\sigma_{\frac{mL}{n}} : m \in \mathbb{N}, \; 0 \le m \le n-1 \right\} = \mathbf{D}_n,$$

where the shift σ_a and the reflection ρ were defined in (4.6). $\mathbf{D}_n$ is the dihedral group of order $2n$, which is the group of symmetries of regular polygons with n sides. We can define the group action directly on elements of the circle S_L^1. For given $x \in S_L^1$, we write

$$\sigma_{\frac{mL}{n}} x = \left(x - \frac{mL}{n}\right) \mod L, \qquad \rho x = (L - x) \mod L$$

For convenience, we introduce the square bracket to denote mod L:

$$[x] = x \mod L.$$

We explicitly classify the orbits of Σ_n on S_L^1.

© The Author(s), under exclusive license to Springer Nature Switzerland AG 2021
A. Buttenschön, T. Hillen, *Non-Local Cell Adhesion Models*, CMS/CAIMS Books in
Mathematics 1, https://doi.org/10.1007/978-3-030-67111-2_5

Lemma 5.1 *Given $x \in S_L^1$, then the orbit of x under Σ_n is*

$$O_n(x) = \left\{ \left[\pm \left(x - \frac{mL}{n} \right) \right], \; m = 0, \ldots, n-1 \right\}.$$

1. If $x = \kappa \frac{L}{n}$ for some $\kappa \in \{0, \ldots, n-1\}$, then

$$O_n(x) = \left\{ 0, \frac{L}{n}, \frac{2L}{n}, \ldots, \frac{(n-1)L}{n} \right\}$$

and $|O_n(x)| = n$.

2. If $x = \kappa \frac{L}{2n}$ for an odd κ, then

$$O_n(x) = \left\{ \frac{L}{2n}, \frac{3L}{2n}, \ldots, \frac{(2n-1)L}{2n} \right\}$$

and $|O_n(x)| = n$.

3. If $x \neq \kappa \frac{L}{2n}$ for any $\kappa \in \{0, \ldots, 2n-1\}$, then

$$O_n(x) = \left\{ x, \left[x - \frac{L}{n} \right], \left[x - \frac{2L}{n} \right], \ldots, \left[x - \frac{(n-1)L}{n} \right], \right.$$
$$\left. [-x], \left[\frac{L}{n} - x \right], \ldots, \left[\frac{(n-1)L}{n} - x \right] \right\}$$

and $|O_n(x)| = 2n$.

Proof We check when two elements of the orbit are identical. Given $m, k \in \{0, \ldots, n-1\}$, then

$$\left[x - \frac{mL}{n} \right] = \left[x - \frac{kL}{n} \right]$$

if and only if $m = k$. Moreover,

$$\left[x - \frac{mL}{n} \right] = \left[\frac{kL}{n} - x \right]$$

if there is a $\theta \in \mathbb{Z}$ such that

$$x - \frac{mL}{n} = \frac{kL}{n} - x + \theta L,$$

which implies

$$x = (k + m + \theta n) \frac{L}{2n},$$

i.e.,

$$[x] = \kappa \frac{L}{2n}, \qquad \kappa \in \{0, \ldots, 2n-1\}.$$

This shows that multiples of $\frac{L}{n}$ and multiples of $\frac{L}{2n}$ form their own classes of orbits of lengths n. All other orbits have length $2n$. □

Using the isotropy subgroup, we define the fixed-point subspaces (containing all the functions invariant under actions of the isotropy subgroup).

$$H^2_{\Sigma_n} = \{u \in H^2 : \sigma u = u, \ \forall \sigma \in \Sigma_n\},$$
$$L^2_{\Sigma_n} = \{u \in L^2 : \sigma u = u, \ \forall \sigma \in \Sigma_n\}.$$

Note that both of the above spaces are again Banach spaces, since both are closed subspaces, which follows from the fact that Σ_n is a topological group and hence the action generated by $\sigma \in \Sigma_n$ is continuous. Using these symmetries, we make the observation.

Lemma 5.2 *Given* $u \in H^2_{\Sigma_n}$ *and* $x \in S^1_L$, *then*

$$u(x) = u(\tilde{x}) \quad \text{for all} \quad \tilde{x} \in O_n(x).$$

For each $n \in \mathbb{N}$, we obtain a Σ_n reduced problem of $\mathcal{F}$ such that

$$\mathcal{F} : \mathbb{R} \times H^2_{\Sigma_n} \mapsto L^2_{\Sigma_n} \times \mathbb{R}$$

since, whenever $u \in H^2_{\Sigma_n}$, we have that

$$\mathcal{F}[\alpha, u] = \mathcal{F}[\alpha, \sigma u] = \sigma \mathcal{F}[\alpha, u].$$

For each Σ_n, we define the symmetry-preserving steady-state problem as

$$\mathcal{F}[\alpha, u] = 0, \ u \in H^2_{\Sigma_n}. \tag{5.2}$$

Then this problem (5.2) has bifurcation points as multiples of n, i.e., at $\alpha_{kn}, 1 \le k$. Based on the symmetry, we can identify zeros of the local and non-local derivatives:

Lemma 5.3 *Suppose that* $u \in H^2_{\Sigma_n}$; *then*

$$\mathcal{K}[u]\left(\frac{mL}{2n}\right) = 0, \quad u'\left(\frac{mL}{2n}\right) = 0, \quad \mathcal{K}[u]''\left(\frac{mL}{2n}\right) = 0, \quad 0 \le m \le 2n - 1.$$

Proof From Lemma 4.4, we have that for each $a > 0$ and $\rho_a u(x) = u(2a - x)$,

$$\mathcal{K}[\rho_a u] = -\rho_a \mathcal{K}[u].$$

Since $u \in H^2_{\Sigma_n}$, we have that $\rho_{\frac{mL}{2n}} u = u$, and so we find that

$$\mathcal{K}[u](x) = \mathcal{K}[\rho_{\frac{mL}{2n}} u] = -\rho_{\frac{mL}{2n}} \mathcal{K}[u](x) = -\mathcal{K}[u]\left(\frac{mL}{n} - x\right).$$

Letting $x = \frac{mL}{2n}$, we obtain that

$$\mathcal{K}[u]\left(\frac{mL}{2n}\right) = -\mathcal{K}[u]\left(\frac{mL}{2n}\right).$$

Then from Lemma 3.12, it follows that also

$$u'\left(\frac{mL}{2n}\right) = 0.$$

Similarly for the second derivative of the non-local term $\mathcal{K}[u]''$, we find

$$\mathcal{K}[u]''\left(\frac{mL}{2n}\right) = -\mathcal{K}[u]''\left(\frac{mL}{2n}\right),$$

and thus we must have that $\mathcal{K}[u]''\left(\frac{mL}{2n}\right) = 0$.   □

The ordering of minima and maxima imposed on Σ_n by the dihedral group motivates the following definition of a *tiling* of the domain S_L^1. Intuitively, the tiling segregates the domain into pieces on which the function $u(x)$ is increasing and decreasing.

Definition 5.1 (Domain Tiling) For $n \in \mathbb{N}$, we define a *tile* by

$$T_i := \left(\frac{iL}{2n}, \frac{(i+1)L}{2n}\right), \; i = 0, \ldots, 2n - 1.$$

The collection $\mathbf{T}_n := \{T_i\}_{i=0}^{2n-1}$ is called the *tiling* of S_L^1, (see Fig. 5.1) and we have that

$$S_L^1 = \mathrm{cl}\left(\bigcup_{i=0}^{2n-1} T_i\right).$$

The collection of tile boundaries is denoted by

$$\partial\mathbf{T}^n = S_L^1 \setminus \bigcup_{i=0}^{2n-1} T_i.$$

A key motivation for this tiling is the fact that each tile contains exactly one point of a canonical orbit.

Lemma 5.4 *Consider* $x \in S_L^1$.

1. *If* $x \neq \kappa\frac{L}{2n}$ *for any* $\kappa \in \{0, 2n - 1\}$, *then each tile* T_i *contains exactly one element of the orbit* $O_n(x)$.
2. *If* $x = \kappa\frac{L}{2n}$ *for some* $\kappa \in \{0, 2n - 1\}$, *then*

$$O_n(x) \subset \partial\mathbf{T}^n.$$

Proof The second case is obvious; hence, we consider $x \neq \kappa \frac{L}{2n}$. Note that we have $2n$ tiles and $|O_n(x)| = 2n$; thus, if we can show that $T_i \cap O_n(x) \neq \emptyset$ for all $i = 0, \ldots 2n-1$, then we are done. Assume there is an index j such that

$$T_j \cap O_n(x) = \emptyset. \tag{5.3}$$

If j is even, then, by symmetry, this implies that the orbit $O_n(x)$ does not intersect any even tile. If j is odd, then this means that $O_n(x)$ does not intersect any odd tile. In either case, we know that for $x \in T_0$, we have $\rho x = L - x \in T_{2n-1}$. While T_0 is an even tile, T_{2n-1} is an odd tile. Consequently, the orbit does not have an intersection with any tile, which means that it has to contain only boundary points. But this is case 2. Hence, the assumption (5.3) is false. □

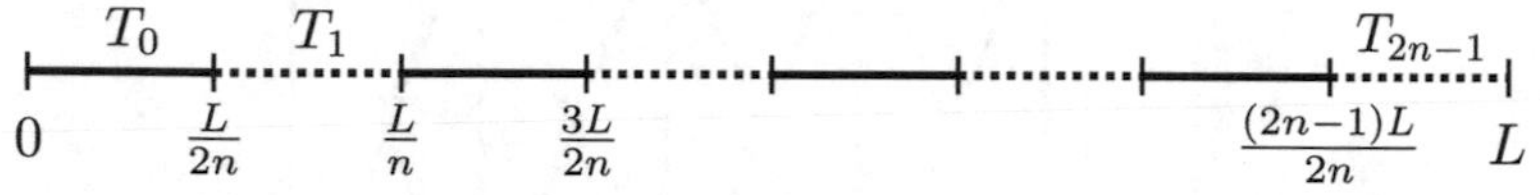

Fig. 5.1 A tiling for S_L^1 as defined in Definition 5.1 and where $n \in \mathbb{N}$. The solid components form T^1, while the dashed components form T^2 (see Definition 5.2). $\partial \mathbf{T}$ is denoted by the vertical lines

Motivated by Lemma 5.3, we use the tiling to define function spaces of functions which have alternating regions on which they are increasing and decreasing.

Definition 5.2 (Spaces of Spiky Functions) Let $n \in \mathbb{N}$, and let $\mathbf{T}_n$ be the tiling of S_L^1 as defined in Definition 5.1. We define two additional collections of tiles

$$T^1 := \bigcup_{i=0}^{n-1} T_{2i}, \qquad T^2 := \bigcup_{i=0}^{n-1} T_{2i+1}.$$

For each $n \in \mathbb{N}$, we define two spaces of spiky functions to be, the space of functions whose derivative has $2n$ simple zeros located on $\partial \mathbf{T}_n$

$$S_n^+ = \left\{ u \in C^2 : u' > 0 \text{ in } T^1,\ u' < 0 \text{ in } T^2,\ u''(x) \neq 0,\ x \in \partial \mathbf{T}_n \right\},$$

and

$$S_n^- = \left\{ u \in C^2 : -u \in S_n^+ \right\}.$$

Remark 5.1 It is easy to see that both $S_n^\pm$ are nonempty since $\cos\left(\frac{2\pi n x}{L}\right) \in S_n^-$, while the negative of this function is an element of S_n^+ (see Fig. 5.2).

Lemma 5.5 *Let* $u \in H^2_{\Sigma_n}$ *be a positive solution of Eq. (5.2) and* $m \in \{0, \ldots, n\}$ *a natural number. Then we have*

$$\frac{1}{L} \int_0^{\frac{mL}{n}} u(x)\,\mathrm{d}x = \bar{u}\frac{m}{n}, \quad and \quad \frac{1}{L} \int_0^{\frac{L}{2n}} u(x)\,\mathrm{d}x = \frac{\bar{u}}{2n}.$$

Proof Let $u \in H^2_{\Sigma_n}$ be a positive solution of Eq. (5.2); then

$$\int_0^{\frac{mL}{n}} u(y)\,\mathrm{d}y = \sum_{i=0}^{m-1} \int_{i\frac{L}{n}}^{(i+1)\frac{L}{n}} u(x)\,\mathrm{d}x = \bar{u}\frac{mL}{n},$$

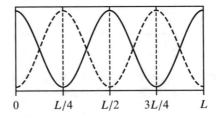

$$0 \qquad L/4 \qquad L/2 \qquad 3L/4 \qquad L$$

Fig. 5.2 The solid black line is an example of a function in S_2^-, while the dashed line is a function in S_2^+

since u is $\frac{L}{n}$-periodic. Furthermore

$$L\bar{u} = \int_0^{\frac{L}{2n}} u(x)\,\mathrm{d}x + \sum_{i=2}^{2n} \int_{\frac{i-1}{2n}L}^{\frac{i}{2n}L} u(x)\,\mathrm{d}x.$$

For the individual integrals, we write

$$\int_{\frac{i-1}{2n}L}^{\frac{i}{2n}L} u(y)\,\mathrm{d}y = \int_0^{\frac{L}{2n}} u\left(x + \frac{i-1}{2n}L\right)\,\mathrm{d}x.$$

If i is an odd integer, then $i - 1$ is even, and for m large enough, we find

$$u\left(x + \frac{i-1}{2n}L\right) = \sigma_{\frac{mL}{n}}\, u\left(x + \frac{i-1}{2n}L\right)$$

$$= u\left(x - \frac{2m - i + 1}{2n}L\right)$$

$$= u\left(x - \frac{M}{n}L\right)$$

$$= \sigma_{\frac{ML}{n}}\, u(x)$$

$$= u(x)$$

for $M = \frac{1}{2}(2m - i + 1)$. Hence

$$\int_{\frac{i-1}{2n}L}^{\frac{i}{2n}L} u(y)\, dy = \int_0^{\frac{L}{2n}} u(x)\, dx.$$

If i is even, we have

$$u\left(x + \frac{i-1}{2n}L\right) = u\left(x - \frac{1}{2n}L + \frac{i}{2n}L\right)$$

$$= \sigma_{\frac{i/2}{n}} u\left(x - \frac{L}{2n}\right)$$

$$= u\left(x - \frac{L}{2n}\right) \qquad\qquad \square$$

Again

$$\int_{\frac{i-1}{2n}L}^{\frac{i}{2n}L} u(y)\, dy = \int_0^{\frac{L}{2n}} u\left(x - \frac{L}{2n}\right) dx = \int_0^{\frac{L}{2n}} u(x)\, dx.$$

We obtain that

$$L\bar{u} = 2n \int_0^{\frac{L}{2n}} u(x)\, dx.$$

5.1 An Area Function

In this section, we explore a type of non-local convexity that is exhibited by an area function of solutions $u(x)$ of Eq. (4.1a). The area function will be defined below in Definition 5.3, and we will see that it is intimately tied to the non-local operator $\mathcal{K}[u]$. In fact, in Proposition 5.1, we use this area function to prove a non-local maximum principle.

Definition 5.3 Let $u \in H^2_{\Sigma_n}$ be a solution of Eq. (4.1); then its *modified area function* is defined to be

$$w(x) := \int_0^x u(y)\, dy - \bar{u}x.$$

where $\bar{u}$ is defined in Eq. (3.2). An example of $u(x)$ and the corresponding $w(x)$ is shown in Fig. 5.3.

Lemma 5.6 *Let* $u(x) \in H^2_{\Sigma_n}$ *be a positive solution of Eq. (4.1), and let* $\mathbf{T}_n$ *be a tiling of* S^1_L. *Then the function* $w(x)$ *defined in Definition 5.3 has the properties:*

1. $w(x)$ *is periodic with period* $\frac{L}{n}$ *and* $w \in C^2$.
2. $w(x)$ *has the following symmetry property*

$$w(x) = -w\left(\frac{mL}{n} - x\right), \ m \in \{0, \dots, n-1\}.$$

From this, it follows that $w(x) = 0$ *for* $x \in \partial \mathbf{T}_n$.

3. *If* $h(u) = u$ *and* $\omega(r) \equiv {}^1/_2$, *then*

$$\mathcal{K}[u](x) = \Delta_1 w(x) := \frac{1}{2}\left[w(x+1) + w(x-1) - 2w(x)\right].$$

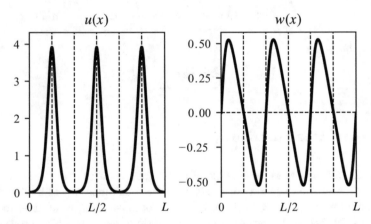

Fig. 5.3 Left: a typical three-peaked solution of Eq. (4.2). Right: the function $w(x)$ as defined in Definition 5.3 corresponding to the solution $u(x)$ on the left

Proof

1. Since $u \in H^2 \subset C^1$, we have that $w \in C^2$. Since $u(x) \in H^2_{\Sigma_n}$, we have

$$w\left(x + \frac{L}{n}\right) = \int_0^{x+\frac{L}{n}} u(y)\,\mathrm{d}y - \bar{u}\left(x + \frac{L}{n}\right)$$

$$= \int_0^x u(y)\,\mathrm{d}y - \bar{u}x + \left(\int_0^{\frac{L}{n}} u(y)\,\mathrm{d}y - \bar{u}\frac{L}{n}\right)$$

$$= w(x),$$

where we use Lemma 5.5. We observe that $w(0) = w(L) = 0$ and $w'(0) = w'(L)$, so that $w(x)$ is periodic.

2. We note that if $u \in H^2_{\Sigma_n}$, we have from Lemma 5.5 that

$$\int_0^x u(y)\,dy = L\bar{u}\frac{m}{n} - \int_0^{\frac{mL}{n}-x} u(y)\,dy.$$

Using this identity, it's easy to verify that $w(x) + w\left(\frac{mL}{n} - x\right) = 0$.
3. In this case, $h(u) = u$ is linear, and the linear factors cancel out such that only the integral terms of $w(x)$ remain. □

Remark 5.2 Lemma 5.6 (3) makes a strong connection between the area function w and the non-local operator $\mathcal{K}$ in the linear case. Such a connection is not so easily obtained if h is more general. In that case, we would define

$$w(x) := \int_0^x h(u(y))\,dy - \overline{h(u)}x$$

and the relations to $\mathcal{K}$ are less clear. Hence, in the following, we restrict to the linear case of $h(u) = u$.

5.1.1 A Non-local Maximum Principle

We assume $h(u) = u$.

Lemma 5.7 *Let $n \in \mathbb{N}$, $\mathbf{T}_n$ be a tiling of S^1_L, and*

$$u \in \left\{ v \in C^1 : v' \geq 0 \text{ in } T^1,\ v' \leq 0 \text{ in } T^2 \right\},$$

where the T^i are defined in Definition 5.1. Let $T \in \mathbf{T}_n$ be any tile; then

1. *If $u'(x) \equiv 0$ in T, then $w(x) \equiv 0$ on T.*
2. *If $u'(\tilde{x}) \neq 0$ for some $\tilde{x} \in T$, then $u'(\tilde{x}) \geq 0$ implies $w(x) < 0$ in T, and $u'(\tilde{x}) \leq 0$ implies $w(x) > 0$ in T.*

Proof Choose any tile $T \subset \mathbf{T}_n$. We assume the tile has been chosen such that $u'(x) \leq 0$. The alternative case follows a similar argument. Note that on ∂T we have due to Lemma 5.6 that $w = 0$. Then there are two cases to consider:

1. If $u'(x) \equiv 0$, we find that $w''(x) = u'(x) \equiv 0$ and w satisfies the boundary value problem

$$\begin{cases} w''(x) = 0 & \text{in } T \\ w = 0 & \text{on } \partial T, \end{cases}$$

hence, by direct integration, we find that $w \equiv 0$.

2. If $u'(x) \le 0$ in T and $u'(\tilde{x}) < 0$ for some $\tilde{x} \in T$, then $w''(\tilde{x}) = u'(\tilde{x}) < 0$, and w satisfies the boundary value problem

$$\begin{cases} w''(x) \le 0 & \text{in } T \\ \quad\; w = 0 & \text{on } \partial T. \end{cases}$$

Then by the elliptic minimum principle (Theorem 1.4 from [117]), w attains its minimum on the boundaries, and $w(x) > 0$ in T. Since w is differentiable, we also have that

$$\frac{\partial w}{\partial \mathbf{n}} < 0,$$

where $\mathbf{n}$ is the unit outward normal at ∂T.

If $u'(x) \ge 0$ on T and $u'(\tilde{x}) > 0$ for some $\tilde{x} \in T$, then the elliptic maximum principle shows that $w < 0$ on T and

$$\frac{\partial w}{\partial \mathbf{n}} > 0,$$

where $\mathbf{n}$ is the unit outward normal at ∂T.   □

On a given tile T, we now combine the concavity of w and the non-locality of $\mathcal{K}[u]$ to obtain a type of non-local maximum principle for the solutions of the steady-state equations (4.1). Key in establishing this result are the symmetries of $H^2_{\Sigma_n}$, which allow us to restrict the non-local operator $\mathcal{K}[u]$ to the canonical tile $T = (0, \frac{L}{2n})$. In more detail, any time the sensing domain (refer to Definition 3.1) of $\mathcal{K}[u]$ reaches a tile boundary, it is reflected. The precise final locations of the sensing domain end points are important in what follows. This is formalized by introducing the reflection (subsequent definition) and its properties (subsequent Lemma).

Let $T_0 \in \mathbf{T}_n$ be a given tile, and consider $x \in T_0$. The sensing domain of the non-local operator is

$$E(x) = [x - 1, x + 1] \subset S^1_L,$$

understood as a closed interval on the circle. Since we are in a symmetric domain, and since the domain length $L > 2$, we can assume without loss of generality that $0 < x - 1 < L$ and $0 < x + 1 < L$, such that we do not need to worry about the domain wrapping around the boundary of the parameterization on $[0, L]$.

The end points of integration $x - 1$ and $x + 1$ might reach into neighboring tiles; hence, we consider the orbits of these points

$$O_n(x - 1) \qquad \text{and} \qquad O_n(x + 1).$$

We had shown in Lemma 5.4 that there are three types of orbits: those inside the tiles, which have length $2n$, those on even boundary points $\theta \frac{L}{n}$, and those residing on odd boundary points $(2l - 1)\frac{L}{2n}$. If $x - 1$ and $x + 1$ are not in $\partial \mathbf{T}_n$, then each orbit has exactly one representative in our starting tile T_0, and we can compute this representative explicitly.

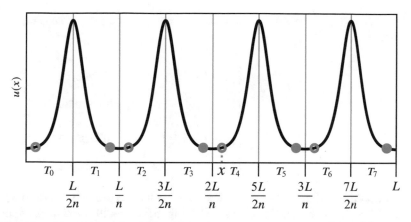

Fig. 5.4 The orbits of x as defined in Definition 5.4. The orbits of the form $\left[x - \theta\frac{L}{n}\right]$ denoted by open circles ($\circ$) and $\left[\theta\frac{L}{n} - x\right]$ denoted by filled circles ($\bullet$)

Definition 5.4 Given a tile T_0, for $x \notin \partial\mathbf{T}_n$, we denote the representative of x in T_0 as $R(x)$.

The points on the orbits can be written in one of two ways,

$$\text{either}\qquad \left[x - \theta\frac{L}{n}\right]\qquad \text{or}\qquad \left[\theta\frac{L}{n} - x\right],\qquad \theta \in \{0,\ldots,n-1\}$$

where we again use the square brackets to indicate mod L (see Fig. 5.4). We can compute the representative as follows.

Lemma 5.8 *Consider a given tile T_0 and some $x \notin \partial\mathbf{T}_n$; then there exists a unique index $l \in \{0,\ldots,2n-1\}$ such that*

$$\text{either } R(x) = \left[x - l\frac{L}{2n}\right]\qquad \text{and}\qquad l \text{ is even,}$$

$$\text{or } R(x) = \left[(l-1)\frac{L}{2n} - x\right]\qquad \text{and}\qquad l \text{ is odd.}$$

Proof If $R(x)$ has the form of $\left[x - \theta\frac{L}{n}\right]$, then we chose an even l and write

$$R(x) = \left[x - \theta\frac{L}{n}\right] = \left[x - l\frac{L}{2n}\right].$$

If $R(x)$ is of the form $\left[\theta\frac{L}{n} - x\right]$, then we chose an odd l and write

$$R(x) = \left[\theta\frac{L}{n} - x\right] = \left[(l-1)\frac{L}{2n} - x\right].$$

As the representative is unique, only one of these cases can arise. $\square$

Then we get explicit equations for the representation of the sensing domain end points

$$R(x-1) = \begin{cases} [x - 1 - l\frac{L}{2n}] & \text{if } l \text{ is even} \\ [(l-1)\frac{L}{2n} - (x-1)] & \text{if } l \text{ is odd.} \end{cases} \tag{5.4}$$

$$R(x+1) = \begin{cases} [x + 1 - k\frac{L}{2n}] & \text{if } k \text{ is even} \\ [(k-1)\frac{L}{2n} - (x+1)] & \text{if } k \text{ is odd.} \end{cases} \tag{5.5}$$

The corresponding numbers of reflections l and k are denoted as l_x and k_x, where x is the center of the sensing domain. See Fig. 5.5 for an illustrative example.

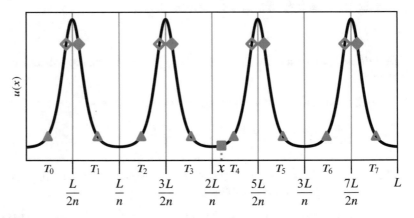

Fig. 5.5 The two different orbits of $R(x-1)$ denoted by open and closed diamonds ($\diamond$) and $R(x+1)$ denoted by open and closed triangles ($\triangle$). The location of x is denoted by a filled square ($\square$). Key is that each tile T_i contains a representative of each of $R(x-1)$ and $R(x+1)$

Lemma 5.9 *Let $x \in T$, $x \notin \partial T_n$ and let $R(x-1)$, $R(x+1)$ denote the representatives of the left and right integral boundary points in T. Let l_x denote the corresponding index for $R(x-1)$ and r_x the index for $R(x+1)$, as defined in (5.4, 5.5).*

1. Then $|x - R(x+1)| \le 1$ and $|x - R(x-1)| \le 1$, where in both cases equality is achieved if and only if $k_x, l_x = 0$.
2. Moreover $|l_x - k_x| \le 1$.
3. $w(x-1) = (-1)^{l_x} w(R(x-1))$, $w(x+1) = (-1)^{k_x} w(R(x+1))$
4. If $h(u) = u$, then

$$\mathcal{K}[u] = \frac{1}{2}\left((-1)^{l_x} w(R(x-1)) + (-1)^{k_x} w(R(x+1)) - 2w(x) \right) \tag{5.6}$$

Proof

1. Since $x - 1$ and $x + 1$ are the integration end points of $\mathcal{K}[u]$, they are separated from x by at most one sensing radius, i.e., $|R(x + 1) - x|$, $|R(x - 1) - x| \leq 1$. Finally, note that equality is only achieved when $k_x, l_x = 0$, since if $k_x \neq 0$ or $l_x \neq 0$, we shift the point $x - 1$ or $x + 1$ closer to x by units of $\frac{L}{n}$, hence reducing the distance.
2. The number of tiles covered to the right or left cannot vary by more than one, since the sensing domain is symmetric and the tiles have uniform length.
3. As we have seen earlier, $w(x + \frac{L}{2n}) = -w(x)$; hence, a shift by increments of $\frac{L}{2n}$ leads to a sign change in w. Hence

$$w(x - 1) = (-1)^{l_x} w(R(x - 1)), \quad w(x + 1) = (-1)^{k_x} w(R(x + 1)).$$

4. Equation (5.6) follows directly from the previous item and Lemma 5.6-3. □

Proposition 5.1 *Let $L > 2$, and $\omega(r) \equiv {}^1/_2$, and $n \in \mathbb{N}$ be such that $M_n(\omega) > 0$. Let $\mathbf{T}_n$ be a tiling of S_L^1. Let $u \in H_{\Sigma_n}^2$ be a positive solution of Eq. (5.2). Further suppose that $h(u) = u$; then $\mathcal{K}[u](x) \neq 0$ for all $x \in \text{int}\, T_i$ for any $T_i \in \mathbf{T}_n$.*

For the proof of Proposition 5.1, we require the following additional lemmas, which we state and prove now.

Lemma 5.10 *Let $u(x) \in H^2$ be a positive solution of Eq. (4.1), and let $w(x)$ be its modified area function defined Definition 5.3. Then if $w(x)$ is linear on $[a, b]$, then $\Delta_1 w \equiv 0$ on $[a, b]$.*

Proof Since $w(x)$ is linear, we have that $w'' = 0$. Since $w(x)$ is the modified area function of a solution of Eq. (4.1), we have that $u' = 0$. By Lemma 3.14, we must have that $\Delta_1 w = 0$ on $[a, b]$ as well. □

The next lemma recalls a useful property of concave functions.

Lemma 5.11 *Let $f \in C^2[a, b]$ be concave, and let $a \leq x_1 < x_2 \leq x_3 < x_4 \leq b$. If the function $f(x)$ satisfies*

$$\frac{f(x_2) - f(x_1)}{x_2 - x_1} = \frac{f(x_4) - f(x_3)}{x_4 - x_3} = M \in \mathbb{R},$$

then $f(x)$ must be linear on $[x_1, x_4]$. If in addition $f(x_1) = f(x_4)$, then $f(x)$ must be constant on $[x_1, x_4]$.

Proof Let $y \in (x_1, x_2)$; then by the concavity of $f(x)$, we have

$$M = \frac{f(x_2) - f(x_1)}{x_2 - x_1} \geq \frac{f(x_2) - f(y)}{x_2 - y} \geq \frac{f(x_4) - f(x_3)}{x_4 - x_3} = M.$$

Rearranging we find that

$$f(y) = f(x_2) - M(x_2 - y) = f(x_1) + M(y - x_1), \quad \text{for } y \in (x_1, x_2).$$

We repeat the same argument for any $y \in (x_2, x_3)$, and $y \in (x_3, x_4)$, to find that

$$f(y) = f(x_1) + M(y - x_1), \quad \text{for } y \in (x_1, x_4).$$

If $f(x_1) = f(x_4)$, then we have

$$f(x_4) = f(x_1) + M(x_4 - x_1) = f(x_1);$$

thus, $M = 0$, and $f(y) = f(x_1) = f(x_4)$.   □

Lemma 5.12 *Let $L > 2$, and $\omega(r) \equiv 1/2$, and $n \in \mathbb{N}$ be such that $M_n(\omega) > 0$ and $h(u) = u$. Let $\mathbf{T}_n$ be a tiling of S_L^1, and denote the canonical tile by T. Let $u \in H_{\Sigma_n}^2$ be a positive solution of Eq. (5.2); then the set*

$$N := \{x \in T : \Delta_1 w(x) = 0\}$$

is both relatively open and closed in T.

Proof The canonical tile is given by $T = (0, {}^L/_{2n})$. Note that when $w \equiv 0$, the result follows immediately. Thus, we consider $w \neq 0$. Without loss of generality, assume that on T we have that $u' \leq 0$ and thus $w > 0$ by Lemma 5.7.

N **is closed:** We note that N can be written as

$$N = (\Delta_1 w)^{-1} (0).$$

Since $w \in C^2$, we have that $\Delta_1 w$ is continuous, implying that N must be closed.

N **is open:** Let $\tilde{x} \in N$; since $\Delta_1 w(\tilde{x}) = 0$, we have

$$\mathcal{K}[u](x) = \frac{1}{2} \left[(-1)^{k_x} w(R(x+1)(k_x)) + (-1)^{l_x} w(R(x-1)(l_x)) - 2w(\tilde{x}) \right]$$

where $R(x-1)(l_x)$ and $R(x+1)(k_x)$ are the end points of the sensing domain of $\mathcal{K}$, reflected by the appropriate (l_x, k_x)-reflection. For a graphical representation of the setup of this proof, see Fig. 5.6. We consider two cases:

l_x *and* k_x **have equal parity:** Since $|k_x - l_x| \leq 1$, we must have that $l_x = k_x$.

- If k_x is odd, we have that

$$\Delta_1 w(\tilde{x}) = \frac{1}{2}\left[-w(R(x+1)) - w(R(x-1)) - 2w(\tilde{x})\right] = 0.$$

Since either $w > 0$ or $w \equiv 0$ on T, we must have that $w \equiv 0$ on T. Then we have $B(\tilde{x}, \varepsilon/2) \subset N$ for $\varepsilon = \text{dist}(\tilde{x}, \partial T)$.

- If k_x is even, we have that

$$\Delta_1 w(\tilde{x}) = \frac{1}{2}\left[w(R(x+1)) + w(R(x-1)) - 2w(\tilde{x})\right]$$

If $k_x \geq 2$, the separation of $R(x+1)$ and $R(x-1)$ is

$$R(x-1) - R(x+1) = \frac{k_x L - 2n}{n}.$$

The case in which the separation is zero coincides with $R(x-1) = R(x+1) = \tilde{x}$, but this is impossible since $L \neq 2n/k_x$ (since $M_n(\omega) > 0$). It's now straightforward to show that $\tilde{x} - R(x+1) = -(\tilde{x} - R(x-1))$, and therefore we have either $R(x-1) < \tilde{x} < R(x+1)$ or $R(x+1) < \tilde{x} < R(x-1)$. If $\Delta_1 w(\tilde{x}) = 0$, we have

$$\mathbb{R} \ni M := \frac{w(R(x+1)) - w(\tilde{x})}{R(x+1) - \tilde{x}} = \frac{w(\tilde{x}) - w(R(x-1))}{\tilde{x} - R(x-1)}$$

Applying Lemma 5.11, we find that w must be linear on $(R(x-1), R(x+1))$ or $(R(x+1), R(x-1))$; a subsequent application of Lemma 5.10 implies that $\Delta_1 w = 0$ on the same intervals. This means that $B(\tilde{x}, \varepsilon/2) \subset N$, for $\varepsilon = |\tilde{x} - R(x+1)|$.

l_x *and* k_x **have different parity:** Since $|l_x - k_x| \leq 1$, we have from Eqs. (5.4) and (5.5) that

$$R(x-1) - R(x+1) = \begin{cases} 2(-1)^{k_x} (L/2n - x), & \text{if } k_x = l_x + 1 \\ 2(-1)^{k_x+1} x, & \text{if } k_x = l_x - 1 \end{cases}$$

Note that since $x \notin \partial T_n$, we have $R(x-1) - R(x+1) \neq 0$. Without loss of generality, consider the case in which l_x is odd and k_x is even (the inverse case follows by the same argument with a sign flip). In this case,

$$\Delta_1 w(\tilde{x}) = 0 \quad \Longleftrightarrow \quad w(R(x-1)) - w(R(x+1)) = -2w(\tilde{x}).$$

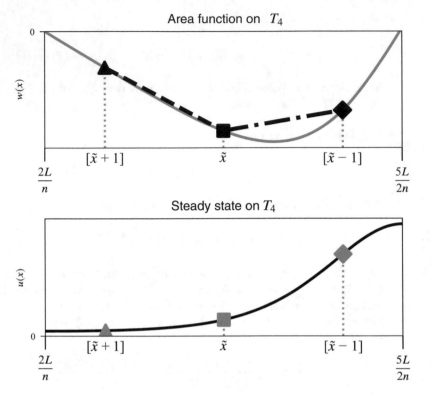

Fig. 5.6 Top: shows the area function $w(x)$ on a selected tile. Bottom: the corresponding spiky function $u(x)$. In both the location of x is indicated by a square, the location of the representative of $[x-1]$ is indicated by a diamond, and the representative of $[x+1]$ is indicated by a triangle. The two secant lines of the area function connecting $([x+1], x)$ and $(x, [x-1])$ are shown by a dashed and dashed dotted line, respectively

- If $k_x = l_x + 1$, then $k_x \geq 2$ and $R(x-1) > R(x+1)$, and since $R(x-1) \leq L_n$ and $R(x-1) - R(x+1) = 2(L/2n - \tilde{x})$, we have $\tilde{x} - R(x+1) \geq L/2n - \tilde{x}$ and $R(x+1) < \tilde{x}$. We must consider two configurations (1) $R(x+1) < R(x-1) < \tilde{x} < L/2n$ and (2) $R(x+1) < \tilde{x} < R(x-1) < L/2n$. For configuration (1), $\Delta_1 w(\tilde{x}) = 0$ is equivalent to

$$\frac{w(R(x-1)) - w(R(x+1))}{2(L/2n - \tilde{x})} = \frac{w(R(x-1)) - w(R(x+1))}{R(x-1) - R(x+1)} = \frac{-2w(\tilde{x})}{2(L/2n - \tilde{x})}.$$

By Lemma 5.11, we have that w is linear on $(R(x+1), L/2n)$. It follows by Lemma 5.10 that $\Delta_1 w = 0$ on the same interval. Then $B(\tilde{x}, {}^{\varepsilon}/_2) \subset N$, for $\varepsilon = L/2n - \tilde{x}$.

For configuration (2), the properties of $R(x-1)$ and $R(x+1)$ and the concavity of w imply that

$$\frac{R(x-1)-\tilde{x}}{\tilde{x}-R(x+1)} \le 1, \qquad \frac{w(\tilde{x})-w(R(x+1))}{\tilde{x}-R(x+1)} \ge \frac{w(R(x-1))-w(\tilde{x})}{R(x-1)-\tilde{x}}.$$

For $R(x-1) < L/2n$, we have $w(R(x-1)) > 0$, resulting in

$$w(\tilde{x})-w(R(x+1)) > -w(R(x-1))-w(\tilde{x}) \quad \Longleftrightarrow \quad \Delta_1 w(\tilde{x}) < 0.$$

If $\Delta_1 w(\tilde{x}) = 0$, we must have that $w \equiv 0$ on $[R(x+1), R(x-1)]$. Then $B(\tilde{x}, {}^{\varepsilon}/_2) \subset N$ for $\varepsilon = R(x-1) - \tilde{x}$.

If $R(x-1) = L/2n$, then $\Delta_1 w(\tilde{x}) = 0$ is equivalent to

$$\frac{w(R(x-1))-w(\tilde{x})}{L/2n-\tilde{x}} = \frac{w(\tilde{x})-w(R(x+1))}{L/2n-\tilde{x}}.$$

Once again, apply Lemma 5.11 and Lemma 5.10 to find that $\Delta_1 w = 0$ on $(R(x+1), L/2n)$. Thus, $B(\tilde{x}, {}^{\varepsilon}/_2) \subset N$, for $\varepsilon = L/2n - \tilde{x}$.

• If $k_x = l_x - 1$, then $R(x-1) < R(x+1)$, and $2x \le R(x+1)$. We once again have to consider two configurations: (1) $x < R(x-1) < R(x+1)$ and (2) $R(x-1) < x < R(x+1)$. For configuration (1), $\Delta_1 w(\tilde{x}) = 0$ is equivalent to:

$$\frac{w(R(x+1))-w(R(x-1))}{R(x+1)-R(x-1)} = \frac{w(R(x+1))-w(R(x-1))}{2x} = \frac{2w(x)-2w(0)}{2x-0}.$$

By Lemma 5.11, we have that w is linear on $(0, R(x+1))$ and $\Delta_1 w = 0$ on the same interval by Lemma 5.7. Then $B(\tilde{x}, {}^{\varepsilon}/_2) \subset N$, for $\varepsilon = \tilde{x}$.

For configuration (2), the properties of $R(x-1), R(x+1)$ and the concavity of w imply that

$$\frac{\tilde{x}-R(x-1)}{\tilde{x}-R(x+1)} \ge -1, \qquad \frac{w(x)-w(R(x-1))}{x-R(x-1)} \ge \frac{w(R(x+1))-w(x)}{R(x+1)-x}.$$

For $R(x-1) > 0$, we have that $w(R(x-1)) > 0$; hence, it follows that

$$w(\tilde{x})+w(R(x-1)) > w(\tilde{x})-w(R(x-1)) \ge w(R(x+1))-w(\tilde{x}).$$

Thus we find that

$$w(R(x+1))-w(R(x-1))-2w(\tilde{x}) < 0 \quad \Longleftrightarrow \quad \Delta_1 w(\tilde{x}) < 0.$$

If $\Delta_1 w(\tilde{x}) = 0$, we must have that $w \equiv 0$ on $(\tilde{x}, R(x+1))$. Then $B(\tilde{x}, {}^{\varepsilon}/_2) \subset N$ for $\varepsilon = \tilde{x} - R(x-1)$.

If $R(x-1) = 0$, $\Delta_1 w(\tilde{x}) = 0$ is equivalent to

$$\frac{w(R(x+1))-w(\tilde{x})}{\tilde{x}} = \frac{w(\tilde{x})-w(0)}{\tilde{x}}.$$

Then we can once again use Lemmas 5.11 and 5.10 to find that $\Delta_1 w = 0$ on $(0, R(x+1))$. Thus $B(\tilde{x}, {}^{\varepsilon}/_2) \subset N$, for $\varepsilon = \tilde{x}$. $\qquad\square$

Proof of Proposition 5.1 Without loss of generality, let $x \in \left(0, {}^{L}/_{2n}\right) =: T$ be the canonical tile on which $w'' \leq 0$. Since u is non-constant, we have that $w > 0$ on T. Suppose that the set

$$N = \{x \in \text{int } T : u'(x) = 0\} \neq \emptyset.$$

Note that by Lemma 3.14, we have that $u'(x) = 0$ if and only if $\Delta_1 w(x) = 0$; thus, the set N here and the set N in Lemma 5.12 coincide. By Lemma 5.12, the set N is both relatively open and closed in T. But T is connected, so that N must equal T. But this means that $u' \equiv 0$ on T, a contradiction. This means that N must be empty. $\quad\square$

5.2 Global Bifurcation Branches for Linear Adhesion Function

Theorem 5.1 *Let $\mathcal{F}$ be given as in Eq. (4.2) with $h(u) = u$, fix $\mathbb{N} \ni n > 0$, assume the assumptions from Theorem 4.1 hold, and let $\bar{u} > 0$ be given. Further let $\Gamma_n = (\alpha_n(s), u_n(x, s))$, $s \in (-\delta_n, \delta_n)$ denote the local bifurcation branch from Theorem 4.1. Then the set of solutions of Eq. (5.2) contains a closed connected set $\mathfrak{C} \subset \mathbb{R} \times H^2_{\Sigma_n}$ such that*

(1) $\mathfrak{C}$ contains $(\alpha_n(s), u_n(x, s))$ for $s \in (-\delta_n, \delta_n)$, where α_n and u_n were defined in (4.16) and (4.17).
(2) For any $(\alpha, u) \in \mathfrak{C}$, we have that $\alpha > 0$, $u > 0$.
(3) $\mathfrak{C} = \mathfrak{C}^+ \cup \mathfrak{C}^-$ can be split into the positive and negative direction of the n-th eigenfunction given in (5.1). $\mathfrak{C}^\pm$ are closed and connected subsets of $\mathfrak{C}$ such that $\mathfrak{C}^+ \cap \mathfrak{C}^- = \{(\alpha_n, \bar{u})\}$. We denote $\mathfrak{C}^\pm_n := \mathfrak{C}^\pm \setminus \{(\alpha_n, \bar{u})\}$, and we have $\mathfrak{C}^+_n \subset \mathbb{R} \times S^-_n$ and that $\mathfrak{C}^-_n \subset \mathbb{R} \times S^+_n$, where $S^\pm_n$ are defined in Definition 5.2.
(4) The unilateral branches $\mathfrak{C}^\pm_n$ are unbounded, that is, for any $\alpha \geq \alpha_n$, there exists $(\alpha, u) \in \mathfrak{C}^\pm_n$.

Proof

1. From Theorem 2.6, it follows that there exists the component $\mathfrak{C}$, which is the maximal, connected, and closed subset of the closure of the set of non-trivial solutions

$$\mathfrak{S} = \left\{ (\alpha, u) \in \mathbb{R} \times H^2_{\Sigma_n} : \mathcal{F}[\alpha, u] = 0, \ u \neq \bar{u} \right\}.$$

   Containing $(\alpha_n, \bar{u})$, this of course is also a consequence of Theorem 4.1. This proves (1).
2. Since $(\alpha_n, \bar{u}) \in \mathfrak{C}$, we have $\alpha_n > 0$. Indeed, if $\alpha \leq 0$, then by the connectedness of $\mathfrak{C}$, there would have to be a point in $\mathfrak{C}$ at which $\alpha = 0$. Thus suppose that $(0, u) \in \mathfrak{C}$, but then from Eq. (4.1a), we have that $u \equiv \bar{u}$. Thus $(0, \bar{u})$ is a bifurcation point. But this contradicts Lemma 4.5 which showed that all bifurcation values are non-zero.

We show the positivity of u by considering

$$\mathcal{P} := \left\{ u \in H^2 : u > 0 \text{ in } S_L^1 \right\}.$$

Then we want to show that $\mathfrak{C} \subset \mathbb{R} \times \mathcal{P}$. We first note that $\bar{u} \in \mathcal{P}$, and by Theorem 4.1, we have that the solution component around the bifurcation point is also in $\mathcal{P}$. Since $\mathfrak{C}$ is connected and $\mathcal{P}$ is open, we have that if $\mathfrak{C} \not\subset \mathbb{R} \times \mathcal{P}$, then there exists a $(\alpha, u) \in \mathbb{R} \times \partial \mathcal{P}$ such that $0 \le u$. First, suppose that there exists $\hat{x} \in S_L^1$ such that $u(\hat{x}) = 0$. Then note that Eq. (4.1a) can be written as

$$\begin{cases} -u'' + a(x)u'(x) + b(x)u(x) = 0 \text{ in } [0, L] \\ \mathcal{B}[u, u'] = 0. \end{cases}$$

where

$$a(x) = \alpha \, \mathcal{K}[u](x) \qquad b(x) = \alpha \, \mathcal{K}[u'](x).$$

Due to Lemma 3.12, we have that both $a(x), b(x) \in C^2(S_L^1)$. Then, the maximum principle for non-negative functions implies that $u \equiv 0$. However, this contradicts the integral constraint in Eq. (4.1), which must hold on $\mathfrak{C}$. Thus we have that $\mathfrak{C} \subset \mathbb{R} \times \mathcal{P}$.

3. Then consider the decomposition of $\mathfrak{C}$ into subcontinua such that $\mathfrak{C}^+ \cap \mathfrak{C}^- = \{(\alpha_n, \bar{u})\}$. Since the cosine is decreasing in the first tile, we claim that $\mathfrak{C}_n^+ \subset \mathbb{R} \times S_n^-$. Since $\mathfrak{C}_n^+$ is a connected topological space, it suffices to show that $\mathfrak{C}_n^+ \cap (\mathbb{R} \times S_n^-)$ is nonempty, relatively open, and relatively closed in $\mathfrak{C}_n^+$. Then we conclude that this space is all of $\mathfrak{C}_n^+$. We split the proof of this claim into three pieces.

(1) To show that $\mathfrak{C}_n^+ \cap (\mathbb{R} \times S_n^-)$ is nonempty, we note that due to the local bifurcation result Theorem 4.1 for small s, the solution along the branch is given by Eq. (4.17) with n; thus, the solution branch is in $\mathfrak{C}_n^+ \cap (\mathbb{R} \times S_n^-)$.

(2) To show that $\mathfrak{C}_n^+ \cap (\mathbb{R} \times S_n^-)$ is relatively open in $\mathfrak{C}_n^+$, we use sequential openness. Let $(\hat{\alpha}, \hat{u}) \in \mathfrak{C}_n^+ \cap (\mathbb{R} \times S_n^-)$, and consider any sequence $(\alpha_k, u_k) \subset \mathfrak{C}_n^+$ convergent to $(\hat{\alpha}, \hat{u})$. Then we are left with showing that the tail of this sequence is contained in $\mathfrak{C}_n^+ \cap (\mathbb{R} \times S_n^-)$. Due to the reflections, we must only consider the canonical tile $T_0 := (0, L/2n)$.

Since $\hat{u} \in S_n^-$, we have that $\hat{u}' < 0$ on T_0 and that $u'' \ne 0$ on ∂T_0. Further, we must have that $\hat{u}'$ is decreasing at the left boundary $(R(x-1) = 0)$ and increasing at the right boundary $(R(x+1) = L/2n)$. Hence, we have that

$$\hat{u}''(R(x-1)) < 0 < \hat{u}''(R(x+1)). \tag{5.7}$$

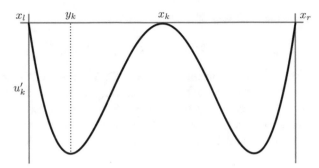

Fig. 5.7 Plot of the derivative u'. The point x_k is such that $u'(x_k) = 0$; this implies the existence of point y_k at which $u''(y_k) = 0$

Since u_k converges to $\hat{u}$, we have that eventually $u'_k \le 0$ on T_0. Suppose that there exists a x_{k_n} (k_n a subsequence of $k \to \infty$) such that $u'(x_{k_n}) = 0$. This implies that there exists y_{k_n} such that $u''_k(y_{k_n}) = 0$ (where either $y_{k_n} \in (R(x-1), x_{k_n})$ or $y_{k_n} \in (x_{k_n}, R(x+1))$).

Without loss of generality, suppose that $y_{k_n} \in (R(x-1), x_{k_n})$ (see Fig. 5.7). Since $\hat{u} \in S_n^-$, we must have that $x_{k_n} \to R(x-1)$ as $k \to \infty$, which implies that $y_{k_n} \to R(x-1)$. Hence, we have that $u''(y_{k_n}) = 0$ for $y_{k_n} \to R(x-1)$ contradicting (5.7).

(3) To show that $\mathbb{C}_n^+ \cap (\mathbb{R} \times S_n^-)$ is relatively closed, we consider the sequence $(\alpha_k, u_k) \subset \mathbb{C}_n^+ \cap (\mathbb{R} \times S_n^-)$ convergent to $(\hat{\alpha}, \hat{u}) \in \mathbb{C}_n^+$. Then again consider T_0 where $u'_k < 0, \forall k$. Then we must have that $\hat{u}' \le 0$. Suppose that there exists $\tilde{x} \in T_0$ such that $\hat{u}'(\tilde{x}) = 0$. This means, since $\hat{u}' \le 0$, that $\hat{u}''(\tilde{x}) = 0$. Evaluating Eq. (3.10) at $\tilde{x}$, we obtain that $\mathcal{K}[\hat{u}](\tilde{x}) = 0$.

If $\tilde{x} \in \text{int } T_0$ and if $\hat{u}$ is non-constant, then Proposition 5.1 implies that $\mathcal{K}[\hat{u}](\tilde{x}) \ne 0$. Thus, we have a contradiction.

If $\tilde{x} \in \partial T_0$ and if $\hat{u}$ is non-constant, then $u'_k(\tilde{x}) = 0, \forall k$, but $u''_k(\tilde{x}) \ne 0, \forall k$. If $\hat{u}''(\tilde{x}) = 0$, Eq. (4.1a) implies that $\mathcal{K}[\hat{u}]'(\tilde{x}) = 0$. Writing $\mathcal{K}[u]$ in terms of $w(x)$ and differentiating, we get that

$$\mathcal{K}[u]'(x) = u(x+1) + u(x-1) - 2u(x).$$

Now since $\hat{u}'(\tilde{x}) = 0$ and $\hat{u}$ is non-constant, thus $\hat{u}$ must either have a global extrema at $\tilde{x}$. Since $L \ne \frac{2n}{k}$, it's impossible for an extrema to be located at $\tilde{x}$ and $\tilde{x} \pm 1$; thus, $\mathcal{K}[\hat{u}]'(\tilde{x})$ cannot be zero.

In the previous, we excluded the possibility that $\hat{u}$ could be constant. Indeed if $\hat{u}$ is constant, then $\hat{u} \equiv \bar{u}$, from the integral constraint in Eq. (4.1). This means that $(\hat{\alpha}, \bar{u})$ is a bifurcation point. Therefore, we must have that $\hat{\alpha} = \alpha_{j_n}$ for some $\mathbb{N} \ni j \ge 1$. The case $j = 1$ cannot occur since $\mathbb{C}_n^+$ does not contain $(\alpha_n, \bar{u})$. For $j \ge 2$, we have from the local bifurcation result (see Theorem 4.1) that in a small neighborhood of $(\alpha_{jn}, \bar{u})$, the solution branch must be given by (4.16). But this means that $u'_k > 0$ and $u'_k < 0$ on T_0, which is a contradiction. Thus, $\mathfrak{C}_n^+ \cap (\mathbb{R} \times S_n^-)$ is closed.

Thus, we have proven that $\mathfrak{C}_n^+ \subset \mathbb{R} \times S_n^-$. For $\mathfrak{C}_n^-$, we argue analogously.

4. Theorem 2.7 implies that each $\mathfrak{C}_n^\pm$ satisfies one of the following alternatives:

   (i) it is not compact in $\mathbb{R} \times H^2_{\Sigma_n}$,
   (ii) it contains a point $(\hat{\alpha}, \bar{u})$ where $\hat{\alpha} \neq \alpha_n$,
   (iii) it contains a point $(\alpha, \bar{u} + \tilde{u})$ where $\tilde{u} \in Y_n \setminus \{0\}$,

where

$$Y_n = \left\{ u \in H^2_{\Sigma_n} : \int_0^L u(x) \cos\left(\frac{2\pi n x}{L}\right) dx = 0 \right\}.$$

If alternative (ii) holds, then $\hat{\alpha}$ is a bifurcation point, which is impossible after the proof of (3).

If alternative (iii) holds, then there is a $(\alpha, \bar{u} + \tilde{u}) \in \mathfrak{C}_n^\pm$ with $\tilde{u} \in Y_n \setminus \{0\}$. This means the following holds, where we integrate by parts

$$0 = \int_0^L \tilde{u}(x) \cos\left(\frac{2\pi n x}{L}\right) dx = -\frac{L}{2\pi n} \int_0^L \tilde{u}'(x) \sin\left(\frac{2\pi n x}{L}\right) dx, \qquad (5.8)$$

and then we note that since $\tilde{u} \in S_n^\pm$, we have that $\sin\left(\frac{2\pi n x}{L}\right)$ and $\tilde{u}'$ have the same zeros. Further, since $\tilde{u}'' \neq 0$ on ∂T, we have that $\tilde{u}'$ must change sign at those points. Hence, we have two cases, either $\sin\left(\frac{2\pi n x}{L}\right) \tilde{u}' \leq 0$ or $\sin\left(\frac{2\pi n x}{L}\right) \tilde{u}' \geq 0$. In both cases, we then find that

$$\int_0^L \tilde{u}'(x) \sin\left(\frac{2\pi n x}{L}\right) dx \neq 0,$$

contradicting equation (5.8). Therefore, only alternative (i) holds, and thus $\mathfrak{C}^\pm$ are non-compact.

Finally, to complete (4), we note that since $\mathfrak{C}^\pm$ are connected, its projection onto the α coordinate are intervals containing α_n. From the a priori estimate for positive solution derived in Lemma 3.13, we find that for bounded α, the solution u is bounded, in particular, uniformly bounded in C^1. From equation,

$$u'' = \alpha u' \mathcal{K}[u] + \alpha u \mathcal{K}[u]'$$

we then also have that u'' is bounded, since u and u' are bounded. Iterating this process one more time for u''', we consider equation,

$$u''' = \alpha u'' \mathcal{K}[u] + 2\alpha u' \mathcal{K}[u]' + \alpha u \mathcal{K}[u]''.$$

The first two terms on the right-hand side are bounded, and the second derivative of $\mathcal{K}[u]$ is also bounded since $u \in C^3$ and $h(\cdot) \in C^2$. Hence, u is bounded in the norm of C^3 and hence H^3 for bounded α. But $H^3 \subset\subset H^2$, and if α were contained in a bounded interval, then $\mathfrak{C}^{\pm}$ would be compact in $\mathbb{R} \times H^2$, which contradicts our earlier observation. Thus the α coordinate must be unbounded. $\qquad\qquad\qquad\qquad\qquad\qquad\qquad\qquad\qquad\qquad\qquad\qquad\qquad\qquad$ $\square$

5.3 Bifurcation Type for Linear Adhesion Function

We study the stability and the type of the first bifurcating branch. Note that this work is limited to the case $h(u) = u$ as we make heavy use of Lemmas 3.6 and 3.8.

For each $n \in \mathbb{N}$, we let $e_n := \cos\left(\frac{2\pi n x}{L}\right)$, which is the function spanning the nullspace of $\mathcal{D}_u \mathcal{F}(\alpha_n, \bar{u})$. Now we decompose the function space H^2 into two pieces

$$H^2 = \operatorname{span}\left\{ \cos\left(\frac{2\pi n x}{L}\right) \right\} \oplus Y_n,$$

where

$$Y_n := \left\{ u \in H^2 : \int_0^L u(x) e_n(x)\, dx = 0 \right\}.$$

Theorem 5.2 *Suppose that all the assumptions of Theorem 5.1 are satisfied and that $M_{2n}(\omega) \neq 2M_n(\omega)$. Then the bifurcation branch Γ_n in a neighborhood of $(\alpha_n, \bar{u})$ is parameterized by*

$$\begin{bmatrix} u_n(x, s) \\ \alpha_n(s) \end{bmatrix} = \begin{bmatrix} \bar{u} \\ \alpha_n \end{bmatrix} + s \begin{bmatrix} \alpha_n \\ 0 \end{bmatrix} \cos\left(\frac{2\pi n x}{L}\right) + s^2 \begin{bmatrix} p_1(x) \\ \alpha_{n,3} \end{bmatrix} + o(s^2)$$

for $s \in (-\delta, \delta)$, where $p_1(x)$ is determined below. Since $\alpha_n{'}(0) = 0$, this is a pitchfork bifurcation, whose direction is determined by the sign of $\alpha_{3,n}$, and its value is given by

$$\alpha_{3,n} = \frac{\alpha_n^3}{4\bar{u}^2} \left[\frac{M_{2n}(\omega) - M_n(\omega)}{M_{2n}(\omega) - 2M_n(\omega)} \right],$$

where $M_n(\omega)$ is defined in Lemma 3.6. The sign of $\alpha_{3,n}$ is thus determined by the properties of the $M_n(\omega)$ function.

Proof From Theorem 2.5, we have that the bifurcating branches are of class C^3 in particular $(\alpha_n(s), u_n(x, s)) \in C^3$ with respect to s. Then we can write an asymptotic expansion of $u_n(x, s), \alpha_n(s)$ for the n-th bifurcation branch (see Theorem 4.1), that is,

$$u_n(s, x) = \bar{u} + s\alpha_n \cos\left(\frac{2\pi n x}{L}\right) + s^2 p_1(x) + s^3 p_2(x) + o(s^3)$$

$$\alpha_n(s) = \alpha_n + s\alpha_{n,2} + s^2 \alpha_{n,3} + o(s^3),$$

where $p_i \in Y_n$. Since $\mathcal{A}[u_n(s,x)] = \bar{u}$, we have that $\mathcal{A}[p_i] = 0$. For the following discussion, we will also need Fourier expressions for the functions $p_i(x)$. That is, we express both by

$$p_i(x) = \sum_{\substack{k=1 \\ k \neq n}}^{\infty} b_k^i \cos\left(\frac{2\pi k x}{L}\right). \tag{5.9}$$

Then we substitute the asymptotic expansion into Eq. (4.1a), and we group the result by associating terms of equal powers of s.

The terms of order $O(s)$ give

$$\alpha_n e_n'' - \alpha_n^2 \bar{u} \mathcal{K}[e_n]' = 0.$$

It is straightforward to verify that this equation is satisfied precisely when α_n is a bifurcation point (see Lemma 4.5).

The terms of order $O(s^2)$ give

$$\underbrace{p_1''(x)}_{\text{I}} - \underbrace{\alpha_n \, \bar{u} \mathcal{K}[p_1]'}_{\text{II}} - \underbrace{\alpha_n^3 \, (e_n \, \mathcal{K}[e_n])'}_{\text{III}} - \underbrace{\alpha_n^2 \alpha_{2,n} \, \bar{u} \mathcal{K}[e_n]'}_{\text{IV}} = 0. \tag{5.10}$$

We will project this equation onto the space spanned by e_n, and we use Roman numerals, (i.e., I), to refer to those projected terms. First we show that $\alpha_{2,n} = 0$. Indeed, by projecting Eq. (5.10) onto the nullspace of $\mathcal{D}_u \mathcal{F}(\alpha_n, \bar{u})$, we will obtain the result. We will show the work term by term in order as they appear in Eq. (5.10). First, since $p_1 \in Y_n$

$$\text{I} = \int_0^L p_1''(x) \cos\left(\frac{2\pi n x}{L}\right) dx = 0,$$

by integration by parts. For the second term, we apply the results of Lemma 3.8 to obtain

$$\mathcal{K}[p_1]' = -\frac{4\pi}{L} \sum_{\substack{k=1 \\ k \neq n}}^{\infty} k b_k^1 M_k(\omega) \cos\left(\frac{2\pi k x}{L}\right).$$

Then projecting those terms on the nullspace easily shows that

$$\text{II} = 0.$$

We proceed similarly for the third term. First, using Lemma 3.6, we obtain

$$(e_n \, \mathcal{K}[e_n])' = \left(\frac{-4\pi n}{L}\right) M_n(\omega) \cos\left(\frac{4\pi n x}{L}\right). \tag{5.11}$$

Then carrying out the projection onto e_n, we obtain

$$\text{III} = \int_0^L (e_n \, \mathcal{K}[e_n])' \cos\left(\frac{2\pi n x}{L}\right) dx = 0.$$

Finally, for the fourth term, using Lemma 3.8, we obtain

$$\text{IV} = \frac{-4\pi n}{L} M_n(\omega) \int_0^L \cos^2\left(\frac{2\pi nx}{L}\right) dx = -2n\pi M_n(\omega).$$

Substituting all these results into Eq. (5.10), we obtain that

$$\alpha_{2,n} = 0.$$

The terms of order $O(s^3)$ are

$$\underbrace{p_2''(x)}_{\text{I}} - \underbrace{\alpha_n \, \bar{u} \mathcal{K}[p_2]'}_{\text{II}} - \underbrace{\alpha_n^2 \, (e_n \, \mathcal{K}[p_1])'}_{\text{III}} - \underbrace{\alpha_n^2 \, (p_1 \, \mathcal{K}[e_n])'}_{\text{IV}} - \underbrace{\alpha_n \alpha_{3,n} \bar{u} \, \mathcal{K}[e_n]'}_{\text{V}} = 0.$$

$$(5.12)$$

Again we project this equation onto the nullspace of $\mathcal{D}_u \, \mathcal{F}(\alpha_n, \bar{u})$ term by term. The first term is

$$\text{I} = \int_0^L p_2''(x) \cos\left(\frac{2\pi nx}{L}\right) dx = 0.$$

The second term, projected onto the nullspace, is given by

$$\text{II} = -\frac{4\pi}{L} \sum_{\substack{k=1 \\ k \neq n}}^{\infty} k b_k^2 M_k(\omega) \int_0^L \cos\left(\frac{2\pi kx}{L}\right) \cos\left(\frac{2\pi nx}{L}\right) dx = 0.$$

The third term is more interesting. Following integration by parts, we obtain

$$\text{III} = -\frac{4\pi n}{L} \sum_{\substack{k=1 \\ k \neq n}}^{\infty} b_k^1 M_k(\omega) \int_0^L \sin\left(\frac{2\pi nx}{L}\right) \cos\left(\frac{2\pi nx}{L}\right) \sin\left(\frac{2\pi kx}{L}\right) dx.$$

In this case, after substituting in the Fourier expansion for p_1 and using Lemma 3.8 to rewrite the non-local term, we encounter integral terms of the form

$$\int_0^L \sin\left(\frac{2\pi nx}{L}\right) \cos\left(\frac{2\pi nx}{L}\right) \sin\left(\frac{2\pi kx}{L}\right) dx = \begin{cases} 0 & \text{if } k \neq 2n \\ L/4 & \text{if } k = 2n. \end{cases}$$

Thus, only the term $k = 2n$ remains from the Fourier expansion, and we obtain

$$\text{III} = -\pi n M_{2n}(\omega) b_{2n}^1$$

The fourth term is similarly. Following integration by parts, we obtain

$$\text{IV} = \frac{2\pi n}{L} \int_0^L p_1(x) \, \mathcal{K}[e_n] \sin\left(\frac{2\pi nx}{L}\right) dx.$$

In this case, after substituting in the Fourier expansion for p_1 and using Lemma 3.8 to rewrite the non-local term, we encounter integrals similar to those for III. Since once again only the term $k = 2n$ remains from the Fourier expansion, we obtain

$$\text{IV} = n\pi M_n(\omega) b_{2n}^1.$$

Finally, we deal with the last term

$$V = \int_0^L \mathcal{K}[e_n]' \cos\left(\frac{2\pi nx}{L}\right) dx$$

$$= -\frac{4\pi n}{L} M_n(\omega) \int_0^L \cos^2\left(\frac{2\pi nx}{L}\right) dx = -2M_n(\omega)\pi n.$$

Substituting each of the projections into Eq. (5.12), we can solve for $\alpha_{3,n}$.

$$\alpha_{3,n} = \frac{\alpha_n}{2\bar{u}} \left(1 - \frac{M_{2n}(\omega)}{M_n(\omega)}\right) \underbrace{\int_0^L p_1(x) \cos\left(\frac{4\pi nx}{L}\right) dx}_{b_{2n}^1}.$$

Thus next we will have to find b_{2n}^1. To find it, we solve the equation of order $O(s^2)$ for $p_1(x)$, and recalling the result from Eq. (5.11), we then obtain

$$p_1''(x) - \alpha_n \bar{u}\mathcal{K}[p_1]' = \alpha_n^3 (e_n \mathcal{K}[e_n])' = -\alpha_n^3 \frac{4\pi n}{L} M_n(\omega) \cos\left(\frac{4\pi nx}{L}\right).$$

Using the Fourier expansion (5.9) of $p_1(x)$ and substituting it into the previous equation and matching modes, we obtain

$$b_{2n}^1 = \frac{1}{2\bar{u}^3} \left(\frac{\pi n}{L}\right)^2 \left[2M_n^2(\omega) - M_n(\omega)M_{2n}(\omega)\right]^{-1}.$$

which we can substitute into our expression for $\alpha_{3,n}$. We find that

$$\alpha_{3,n} = \frac{\alpha_n^3}{4\bar{u}^2} \left[\frac{M_{2n}(\omega) - M_n(\omega)}{M_{2n}(\omega) - 2M_n(\omega)}\right].$$

$\square$

Example 5.1 Supposes that ω is chosen to be the uniform function, i.e., O1. Using the explicit form of $M_n(\omega)$ computed in Example 3.1, we can compute $\alpha_{3,n}$ and, in this case, find that

$$\alpha_{3,n} = \frac{\alpha_n^3}{2\bar{u}^2} \left(1 - \cot^2\left(\frac{\pi n}{L}\right)\right). \tag{5.13}$$

To illustrate the bifurcation structure in (5.13), we compute the bifurcation branches for the choices of $L = 3$ and $L = 5$. In Fig. 5.8, we show the term $(1 - \cot^2(\frac{n\pi}{L}))$

for the two choices of $L = 3, 5$ as a function of n. First we note that for multiples of L, we have no bifurcations. For $L = 3$, all other modes $n = 1, 2, 4, 5, 7, 8$, etc. have a positive value for $\alpha_{n,3}$; hence, the bifurcation is supercritical. For $L = 5$ (Fig. 5.8 on the right), we see that the modes 1, 4, 6, 9, etc. are subcritical, while the modes 2, 3, 7, 8, etc. are supercritical. Two bifurcation diagrams verifying these computations are shown in Figs. 5.9 and 5.10.

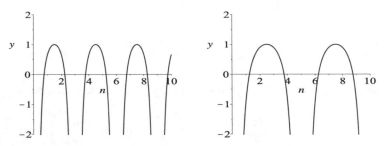

Fig. 5.8 We show the term $1 - \cot^2(\frac{n\pi}{L})$ as a function of n for the two cases of $L = 3$ (left) and $L = 5$ (right). For supercritical modes, the term is positive, and for subcritical modes, it is negative

5.4 Stability of Solutions

So far we have studied the set of solutions of the equation $\mathcal{F}[\alpha, u] = 0$, which are the steady states of the evolution equation

$$\frac{du}{dt} = -\mathcal{F}[\alpha, u]. \tag{5.14}$$

In this section, we are interested in the linear stability of these steady-state solutions. Then the linear stability of such a solution along a branch is determined by the sign of the eigenvalue of $\mathcal{D}_u \mathcal{F}(\alpha(s), u(s))$. An eigenvalue perturbation result proven in [47] shows that the eigenvalue along the trivial solution branch is related to the eigenvalues along the non-trivial solution branch near a bifurcation point. An application of the main result of [47] is the goal of this subsection. Again we limit this to the first solution branch. The eigenvalue problem resulting from linearizing equation (5.14) around solutions in Γ_n near the bifurcation point $(\alpha_n, \bar{u})$ is given by

$$-w'' + \alpha_n(s)(w \, \mathcal{K}[u(s)] + u(s) \, \mathcal{K}[w])' = -\lambda w. \tag{5.15}$$

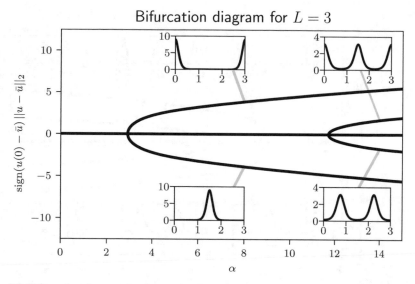

Fig. 5.9 Bifurcation diagram for the non-local adhesion model (4.2) with a uniform integration kernel, when $L = 3$. The bifurcation values and bifurcation types were determined in Example 5.1. In this case, the bifurcations are supercritical pitchfork bifurcations, i.e., we have that $\alpha_{3,1} > 0$ and $\alpha_{3,2} > 0$, and we have a switch of linear stability at the bifurcation point

When $s = 0$, $u(0) = \bar{u}$, we have that

$$-w'' + \alpha_n \bar{u} \left(\mathcal{K}[w]\right)' = -\lambda w. \tag{5.16}$$

Lemma 5.13 *The k-th eigenvalue of the eigenvalue problem in Eq.* (5.16) *is given by*

$$\lambda_k = \begin{cases} 0 & \text{if } n = k \\ \left(\dfrac{2k\pi}{L}\right)^2 \left(\dfrac{n}{k} \dfrac{M_k(\omega)}{M_n(\omega)} - 1\right) & \text{if } n \neq k \end{cases}.$$

Since $M_k(\omega) \to 0$ as $k \to \infty$, we see that $\lambda_k \to -\infty$ as $k \to \infty$.

Proof Apply Lemmas 3.8 and 3.9.   □

Example 5.2 Suppose that ω is chosen to be the uniform function, i.e., O1 (see Section 3.2), and compute λ_k. We find that $\lambda_k > 0$ for $k < n$ and $\lambda_k < 0$ for $k > n$.

To apply the results of [47], we introduce the following definition.

Definition 5.5 (Definition 1.2 [47]) Let $\mathcal{T}, \mathcal{M} \in \mathfrak{L}(X, Y)$. Then $\mu \in \mathbb{R}$ is a $\mathcal{M}$-simple eigenvalue of $\mathcal{T}$ if

$$\dim N[\mathcal{T} - \mu \mathcal{M}] = \operatorname{codim} R[\mathcal{T} - \mu \mathcal{M}] = 1,$$

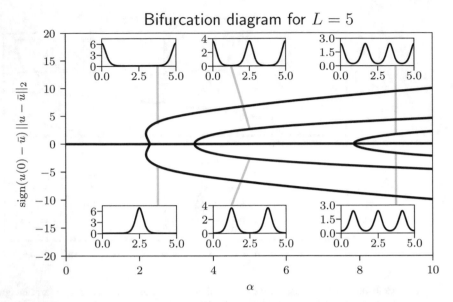

Fig. 5.10 Bifurcation diagram for the non-local adhesion model (4.2) with a uniform integration kernel, when $L = 5$. The bifurcation values and bifurcation types were determined in Example 5.1. In this case, we observe some subcritical pitchfork bifurcations. Indeed, the first, fourth, sixth, ninth, and so on bifurcations will be subcritical pitchfork bifurcations, while the second, third, seventh, eighth, and so on bifurcations are supercritical

and if $N[\mathcal{T} - \mu \mathcal{M}] = \mathrm{span}\{x_0\}$ we have that

$$\mathcal{M}x_0 \notin R[\mathcal{T} - \mu \mathcal{M}].$$

For the purpose here, we define the operator $\mathcal{M} : H^2 \to L_0^2 \times \mathbb{R}$ by

$$\mathcal{M}[w] = \begin{pmatrix} w(x) - \mathcal{A}[w] \\ 0 \end{pmatrix}.$$

Lemma 5.14 $\lambda = 0$ *is a $\mathcal{M}$-simple eigenvalue of $\mathcal{D}_u \mathcal{F}(\alpha_n, \bar{u})$.*

Proof Recall that $\mathcal{D}_u \mathcal{F}(\alpha_n, \bar{u}) : H^2 \mapsto L_0^2 \times \mathbb{R}$ is Fredholm with index 0. Thus, the operator satisfies the first condition in Definition 5.5. We establish the second condition by contradiction. Suppose that

$$\mathcal{M}[e_n] \in R[\mathcal{D}_u \mathcal{F}(\alpha, \bar{u})] = \{u \in L^2 : \mathcal{A}[u] = 0\}.$$

In other words, there exists $w \in H^2$ such that

$$\begin{cases} -w'' + \alpha_n \bar{u} \mathcal{K}[w]' = \alpha_n \cos\left(\frac{2n\pi x}{L}\right) \text{ in } [0, L] \\ \mathcal{A}[w] = 0. \end{cases} \tag{5.17}$$

We expand $w(x)$ using a Fourier series

$$w(x) = \sum_{k=1}^{\infty} w_k \cos\left(\frac{2\pi k x}{L}\right)$$

and substitute that into Eq. (5.17). We then obtain the equation

$$\left(\frac{2k\pi}{L}\right)^2 \left[1 - \frac{n}{k}\frac{M_n(\omega)}{M_k(\omega)}\right] = \begin{cases} 0 & \text{if } n \neq k \\ \alpha_n & \text{if } n = k, \end{cases}$$

which leads to a contradiction when $n = k$. $\qquad\square$

Theorem 5.3 *Let all the assumptions of Theorem 5.1 hold. Then for any $\bar{u} > 0$, and for $s \in (-\delta, 0) \cup (0, \delta)$, the sign of the smallest magnitude eigenvalue of the solution $u(s, x)$ given by Eq. (4.17) of (4.1a) has opposite sign of $\alpha_{3,n}$, in the class of functions such that $\mathcal{A}[u] = \bar{u}$.*

Proof Now we are ready to apply [47, Theorem 1.16]. That implies that there are open intervals I, J with $\alpha_n \in I$, $0 \in J$, chosen such that

$$\gamma : I \mapsto \mathbb{R} \qquad \mu : J \mapsto \mathbb{R},$$

satisfying

$$\gamma(\alpha_n) = \mu(0) = 0,$$

and

$$u : I \mapsto H^2 \qquad w : J \mapsto H^2$$

satisfying

$$u(\alpha_n) = e_n = w(0), \qquad u(\alpha) - e_n \in Y_n, \qquad w(s) - e_n \in Y_n.$$

Then we have two eigenvalue problems

$$\mathcal{D}_u \mathcal{F}(\alpha, \bar{u})[u(\alpha)] = -\gamma(\alpha)\mathcal{M}[u(\alpha)], \quad \text{for } \alpha \in I, \tag{5.18}$$

$$\mathcal{D}_u \mathcal{F}(\alpha(s), u(s))[w(\alpha)] = -\mu(s)\mathcal{M}[w(s)], \quad \text{for } s \in J. \tag{5.19}$$

Whenever $\mu(s) \neq 0$, we have that

$$\lim_{s \to 0, \mu(s) \neq 0} \frac{-s\alpha'(s)\gamma'(\alpha_n)}{\mu(s)} = 1. \tag{5.20}$$

Thus we are left with computing $\gamma'(\alpha_n)$. We differentiate Eq. (5.18) with respect to α and then set $\alpha = \alpha_n$ to obtain

$$\underbrace{-\ddot{u}''}_{\text{I}} + \underbrace{\bar{u}\mathcal{K}[e_n]'}_{\text{II}} + \underbrace{\alpha_n \bar{u}\,\mathcal{K}[\dot{u}]'}_{\text{III}} = \underbrace{-\gamma'(\alpha_n)e_n}_{\text{IV}}.$$

We again project to the nullspace and use Roman numerals to denote the projected terms. We multiply this equation by e_n and integrate by parts. We obtain the following results term-wise. The first term gives us

$$\text{I} = \left(\frac{2n\pi}{L}\right)^2 \int_0^L \dot{u}\cos\left(\frac{2n\pi x}{L}\right)\,dx.$$

The second term gives us, recalling the result from, Lemma 3.8,

$$\text{II} = -2\bar{u}n\pi M_n(\omega).$$

The third term gives us, applying Lemma 3.6, that

$$\text{III} = (\mathcal{K}[\dot{u}]', e_n)_{L^2} = (\dot{u}, \mathcal{K}[e_n]')_{L^2}$$
$$= \frac{-4n\pi}{L}M_n(\omega)\int_0^L \dot{u}\cos\left(\frac{2n\pi x}{L}\right)\,dx.$$

Finally, the last term gives

$$\text{IV} = -\gamma'(\alpha_n)\frac{L}{2}.$$

Combining all the terms, we get

$$\underbrace{\left[\left(\frac{2n\pi}{L}\right)^2 - \alpha_n\bar{u}\frac{4n\pi}{L}M_n(\omega)\right]}_{\text{V}}\int_0^L \dot{u}\cos\left(\frac{2n\pi x}{L}\right)\,dx - 2n\bar{u}\pi M_n(\omega) = -\gamma'(\alpha_n)\frac{L}{2}.$$

Note that when substituting α_n into V, we see that this term is zero, and thus we obtain

$$\gamma'(\alpha_n) = \left(\frac{2n\pi}{L}\right)^2 \frac{1}{\alpha_n}.$$

Substituting γ' and the local expansion for $\alpha(s)$ from Theorem 5.2 into Eq. (5.20), we find that

$$\lim_{s\to 0, \mu(s)\neq 0} -\frac{2}{\alpha_n}\left(\frac{2n\pi s}{L}\right)^2 \frac{\alpha_{3,n}}{\mu(s)} = 1.$$

We conclude that

$$\text{sgn}\,\mu(s) = -\,\text{sgn}\,\alpha_{3,n},$$

for $s \in (0, \delta) \cup (-\delta, 0)$. $\qquad\qquad\qquad\qquad\qquad\qquad\qquad\qquad\square$

5.5 Numerical Verification

In this section, we verify the predictions of Theorem 5.1 on the steady states of Eq. (3.1) by solving this equation numerically. In addition, we also explore numerical solutions in cases not covered by Theorem 5.1 such as non-constant $\omega(r)$ within the non-local term. The code for each figure is available.

5.5.1 Numerical Implementation

The time-dependent equation (3.1) is solved using a method of lines approach, where the spatial derivatives are discretized to yield a large system of time-dependent ODEs (MOL-ODEs). The domain $[0, L]$ is discretized into a cell-centered grid of uniform length $h = 1/N$, where N is the number of grid cells per unit length. Here we set $N = 1024$. The discretization of the advection term utilities a high-order upwind scheme augmented with a flux-limiting scheme to ensure positivity of solutions. For full details on the numerical method, we refer the reader to [74].

The non-local term in Eq. (3.1) presents challenges to its efficient and accurate evaluation. Here we employ the scheme based on the fast Fourier transform introduced in [77]. The resulting system of ordinary differential equations is integrated using the ROWMAP integrator introduced in [188]. Here we use the implementation provided by the authors of [188]. The integrator (written in Fortran) was wrapped using f2py into a SciPy [183] integrate class. The spatial discretization (right-hand side of ODE) is implemented using NumPy [92]. The integrators' error tolerance is set to $v_{tol} = 10^{-6}$.

The bifurcation diagrams in Figs. 5.9 and 5.10 are created by numerical continuation. Here we use pseudo-arclength continuation in the canonical complexification of L^2 to continue known solutions of Eq. (4.2). For the mathematical details on continuation, we refer the reader to [54, 55]. New solution branches are discovered using deflation. In more detail, we use a shifted deflation operator with a H^1 norm, $p = 2$, and $\alpha = 1$; for details, we refer the reader to [64]. The resulting nonlinear equation is solved using the NLEQ-ERR Newton method; for details, see [53]. The required Fréchet derivative is symbolically computed using SymPy [128]. The resulting linear operators are discretized using spectral collocation (using a Fourier basis). For more details on collocating periodic functions, we refer the reader to [178, 190].

---

www.buttenschoen.ca.

http://www.mathematik.uni-halle.de/wissenschaftliches_rechnen/forschung/software/.

https://docs.scipy.org/doc/numpy-dev/f2py/.

5.5.2 Numerical Test Cases

For all numerical solutions in this section, we pick the domain size $L = 5$, and $\bar{u} = 1$ and $\omega \equiv {}^1/_2$. From Lemma 4.5, we know that the first three bifurcation points are located at

$$\alpha_1 = \frac{16\pi^2}{25(5 - \sqrt{5})}, \qquad \alpha_2 = \frac{64\pi^2}{25(5 + \sqrt{5})}, \qquad \alpha_3 = \frac{144\pi^2}{25(5 + \sqrt{5})}.$$

This roughly means that $\alpha_1 \sim 2.28$, $\alpha_2 \sim 3.49$, and $\alpha_3 \sim 7.85$. For all subsequent numerical simulations, we pick a value of α from each of the intervals $(0, \alpha_1)$, (α_1, α_2), and (α_2, α_3). The short-time numerical solutions of Eq. (3.1) are presented in Fig. 5.11. The top row shows the final solution profiles, while the bottom row shows a kymograph with the spatial information on the x-axis and time on the y-axis. These numerical solutions have three key properties: (1) when the value of $\alpha < \alpha_1$, the solution is constant and equals $\bar{u}$; (2) when $\alpha \in (\alpha_1, \alpha_2)$, the solution has a single peak, while when $\alpha \in (\alpha_2, \alpha_3)$ the solution has two peaks; and (3) the peaks are uniformly spaced on the domain. In addition, it is straightforward to check that these solutions are symmetric under the actions of $\mathbf{D}_1$ and $\mathbf{D}_2$, respectively. These observations match the predictions of Theorem 4.1 and Theorem 5.1, and (4) the translational symmetry of Eq. (4.1) is on display, since solution peaks may form anywhere in the domain.

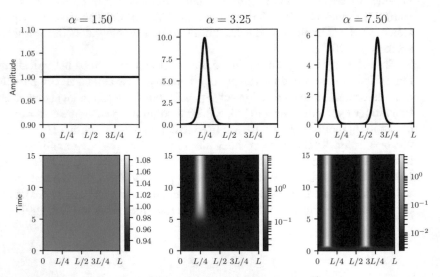

Fig. 5.11 Numerical solutions of Eq. (3.1) on the periodic domain S_L^1. In the top row, we show the final solution profiles, while below are the kymographs. (Left) $\alpha = 1.5$. (Middle) $\alpha = 3.25$. (Right) $\alpha = 7.5$. Here $L = 5$

Next we ask the question whether these patterns persist for a long time. In similar equations, such as models of chemotaxis, it is well known that patterns with many peaks undergo coarsening [147]. Indeed, our numerical solutions of Eq. (3.1) exhibit similar coarsening (see Fig. 5.12). Theorem 5.3 states that near the bifurcation point, solutions with two or more spikes are saddles of Eq. (3.1). While Theorem 5.3 is not valid far away from bifurcation points, the numerical results suggest this to remain true in some cases. Interestingly however, the right kymograph in Fig. 5.12 shows a coarsening from 4 to 2 spikes, without a further coarsening to a single spike (even if the simulation time is increased to $t_f = 10^{13}$).

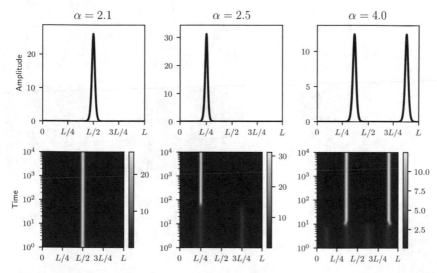

Fig. 5.12 Long-term numerical solutions of Eq. (3.1) on the periodic domain S_L^1. The final solution profiles remain unchanged by either extending the simulation up to 10^{13} or reducing the solver's error tolerance by a factor of 100. (Left) for $\alpha = 2.1$, the single peak that formed initially is stable, while (Middle) for $\alpha = 2.5$, the initial pattern with two peaks coarsens into a singled peaked pattern. (Right) for $\alpha = 4.0$, the initial pattern of four peaks coarsens once to a double peak solution, which does not undergo additional coarsening. Here $L = 10$

While Theorem 5.1 is not valid for non-constant $\omega(r)$, we can use Lemma 4.5 to determine the bifurcation points from the trivial solutions. In three selected cases of non-constant $\omega(r)$, we numerically explore the solutions of Eq. (3.1). The final solution profiles and the corresponding kymographs for $\alpha \in (\alpha_2, \alpha_3)$ are shown in Fig. 5.13. In each case, it is straightforward to verify that the solution profiles are $\mathbf{D}_2$ symmetric and resemble the solutions for constant $\omega(r)$.

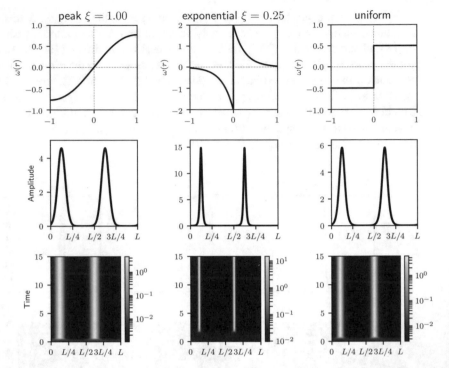

Fig. 5.13 A comparison of solutions for the three different choices of the kernel $\omega(r)$ O1-O3 introduced in Section 3.2. For each simulation $L = 5$, $\alpha = 7.5$, which in each case is greater than α_2, i.e., the second bifurcation point. It was verified by computing $\Delta M_n(\omega)$ that in each case a switch of stability occurs at the bifurcation point α_2. Here $L = 5$

5.6 Summary of the Analytical Challenges

In this chapter, we established the existence of global solution branches of the non-local equation (4.1), where each solution branch originates from the homogeneous solution (see Chap. 4). While it is straightforward to apply the abstract global bifurcation theorem (Theorem 2.7) following the establishment of the local result, the true challenge lies in discerning which of its three alternatives hold. For nonlinear Sturm–Liouville problems, Rabinowitz et al. classified solution branches by the solution's number of zeros. This is based on the fact that its solutions can only have simple zeros, (i.e., $f(x_0) = 0$, but $f'(x_0) \neq 0$). This does not help in our case, since, firstly, we are searching for positive solutions of Eq. (4.1) and, secondly, for second-order elliptic PDEs, the nodal separation is proven by transforming the equation into a two-dimensional initial value problem (IVP). But in the case of non-local equations, this is impossible, since we must look both backward and forward. A different approach to classifying solution branches is to use symmetries; see, for instance, Healey et al. (equivariant nonlinear elliptic equations) [94]. Symmetries have also been used in applications of nonlinear systems of reaction diffusion equations

[72, 73, 140]. Here we show that Eq. (4.1) is $\mathbf{O}(2)$ equivariant. Employing ideas of equivariant bifurcation theory [28, 82, 94], we first collect at each bifurcation point all elements of $\mathbf{O}(2)$ that leave the local nullspace invariant. This leads to the isotropy subgroup, which here is the dihedral group. Next for each isotropy subgroup, we collect all functions, left invariant under its action, in a fixed-point function space. Combining the symmetry properties of the non-local operator $\mathcal{K}$ (Lemma 5.3), and with the particular properties of positive solutions of Eq. (4.1) (Section 3.5), we find that the zeros of the solution's derivative have fixed spatial positions. Using those locations, we introduce the domain tiling Ω_i, on the top of which we construct the spaces of spiky functions $(\mathcal{S}_n^\pm)$. In other words, we classify solution branches by the solution's derivative number of zeros. What remains to be shown is that the branches obtained from the abstract bifurcation theorem Theorem 2.7 are contained in $\mathcal{S}_n^\pm$. The challenge is to ascertain that different branches (with different number of zeros) cannot intersect. In other words, we must show that $u'(x)$ can only have simple zeros. Finally note that due to Lemma 3.14, this question is closely tied to the zeros of the non-local operator $\mathcal{K}[u]$.

Using the equation's symmetries, we reduce the problem to having to show that $u'(x) \neq 0$ in each tile. Classically, for local equations, this is established using a maximum principle (see, for instance, [153, 184, 192]). In the case of uniform integration kernel $\omega(r) = 1/2$, the non-local operator $\mathcal{K}$ and the solution's area function w are related by $\mathcal{K}[u] = \Delta_1 w$. On each tile, the area function $w(x)$ satisfies a Dirichlet problem, and the maximum principle implies that $w > 0$ or $w < 0$. In a novel use of symmetries, we confine the non-local operator $\mathcal{K}[u]$ to a single tile. Then the non-locality of the operator $\mathcal{K}[u]$ establishes that $\mathcal{K}[u] \neq 0$ in any tile, and hence $u'(x) \neq 0$. Due to its similarity to classical maximum principles, we refer to this result as a type of "non-local" maximum principle. This allows us to establish that the branches obtained from the abstract bifurcation theorem (Theorem 2.7) are contained in $\mathcal{S}_n^\pm$ and establish Theorem 5.1.

Using an asymptotic expansion near each bifurcation point, we find that each bifurcation is of pitchfork type. Both supercritical and subcritical bifurcations are possible depending on the sign of $\alpha_{3,n}$ (see Theorem 5.2). Using an eigenvalue perturbation result by Crandall et al. [47], we show that the sign of the smallest magnitude eigenvalue of the linearization about the non-trivial solution is opposite to the sign of $\alpha_{3,n}$. Interestingly, the sign of $\alpha_{3,n}$ is solely determined by the properties of the integration kernel $\omega(r)$ (through $M_n(\omega)$ and $M_{2n}(\omega)$). Two possible bifurcation diagrams are shown in Figs. 5.9 and 5.10.

The global bifurcation result Theorem 5.1 is limited to the linear adhesion model (i.e., $h(u) = u$) and to the uniform integration kernel (i.e., $\omega(r) = 1/2$). Improving both limitations is not easy. Allowing for more general $h(u)$ requires understanding how the area function $w(x)$ is modified (see also Remark 5.2), while allowing for different $\omega(r)$ requires generalizing the "non-local" maximum principle (i.e., result Proposition 5.1). Numerical simulations suggest that for commonly used $\omega(r)$, a similar result should hold (see Fig. 5.13). It is a highly desirable future goal to understand for which integration kernels $\omega(r)$ Theorem 5.1 continues to hold and for which integration kernels the result needs to be modified.

It has to be noted that this result does not exclude the possibility of further secondary bifurcations along the solution branches, which may break further symmetries. The theoretical analysis of secondary bifurcations has remained a challenge. A possible mechanism for identifying further bifurcations is to monitor the Leray-Schauder degree for further sign changes along solution branches. For this task, one has to rely on a combination of numerical exploration and mathematical ingenuity in using the equation's structure to identify secondary bifurcations.

This abstract bifurcation analysis gives rise to several interesting modelling observations. Most noticeable in our analysis were the mathematical properties of the integral kernel $\Omega(r)$ of the non-local operator $\mathcal{K}[u]$. Following the modelling work in [31], this term determines how likely it is that a cell protrusion reaches a particular target. In our analysis, the properties of Ω enter through the quantity $M_n(\omega)$ at several key moments: (1) the sign of $M_n(\omega)$ determines whether or not we have a bifurcation and (2) it determines whether we immediately observe a switch of stability. The minimum adhesion strength (α_1 in Lemma 4.5) which allows the formation of cell aggregates (non-trivial solutions) is reduced with increasing domain size L, magnitude of $M_n(\omega)$, and size of $h'(\bar{u})$, while no parameter increases α_1. Note that these statements only hold for the non-dimensionalized equation.

5.7 Further Reading

Extensions to multiple population systems are highly desirable to study the possible cell sorting patterns. In addition, such extensions promise more intricate dynamics. A different extension could be to consider the model variations proposed by Murakawa et al. [133] and Carrillo et al. [39], who added a density-dependent diffusion term. An extension to higher spatial dimension would be worthwhile, to more realistically study the formation of tissues. In fact, the notion of a tiling in higher dimensions has been considered prior (see, e.g., the work by Courant who defined tilings in higher dimensions [44]). Finally, little is known about the qualitative long-time behavior of solutions, or the structure of its global attractor, for the Armstrong model equation (3.1) and its generalizations.

In [37], Carrillo et al. provide a bifurcation analysis for the n-dimensional Mckean-Vlasov equation (2.16), which includes our model for linear adhesion force function $h(u) = u$. The results presented in this monograph extend the results of Carrillo et al. [37] in the following sense. For the local bifurcation result in one dimension, we allow for general nonlinear adhesion force functions $h(u)$. For the one-dimensional linear case $h(u) = u$, we obtain a global bifurcation result classifying solution branches, and we can identify the types of bifurcations as subcritical and supercritical. Finally, in Part III, we include other, biologically relevant boundary conditions into the analysis.

Spiky-type patterns have been studied extensively in two component reaction diffusion equations with no advection terms. An analytical theory to determine the stability of spike patterns has been developed for the Gierer-Meinhardt [186] and for Gray-Scott models. Similar, developments have been carried out for reaction-diffusion models of species segregation and cross-diffusion [115]. Recently, similar techniques have been extended to include reaction-diffusion equations with chemotactic drift, such as models of urban crime [30, 174, 179], and others have focused on the Keller–Segel model [103, 112, 147, 152, 168]. A highly desirable goal would be to extend these methods to include non-local adhesion models (3.1) and improve upon Theorem 5.3.

A common theme to improve our understanding of scalar and systems of non-local partial differential equation is the synergistic use of numerical and analytical methods. Numerical exploration allows us to explore equations' behavior beyond their current theoretical understanding. Observations made inform the development of novel analytical tools. The properties developed in this chapter, particularly the symmetry properties of solutions, have their origin in numerical solutions of the non-local adhesion model. This requires accurate and precise numerical methods. Since many scalar equations (including the scalar aggregation equation) have a gradient flow structure (see also Sect. 2.6), it is desirable that the used numerical methods are energy dissipating and preserve positivity. This is particularly important to accurately study the asymptotic dynamics and nonlinear stability of stationary solutions. In other words, the scheme has to preserve the gradient flow structure and solution positivity. Such a finite-volume method for nonlinear non-local equations was proposed in [36]. The papers [13, 14] improve their initial work, resulting in a fully discrete energy-dissipating numerical scheme. In [38], this hybrid numerical analytical approach is used to study the asymptotic behavior of a two-species non-local model with cross-diffusion. The insights from the numerical solutions are used to analytically compute stationary steady states and travelling pulse solutions and to study phase transitions as the strength of the cross-diffusivity is varied.

Global bifurcation results for the first solution branches are available for the local Keller–Segel model [184] and non-local chemotaxis model [192]. In [35], the authors study phase transitions resulting from variations in the chemosensitivity parameter of a Keller–Segel model with volume exclusion on a one-dimensional domain with Neumann boundary conditions. To identify bifurcation points, the local result of Crandall-Rabinowitz is applied (as done in the previous chapter). Unfortunately the global version cannot be applied, since the solution $u = 0$ is degenerate. Nevertheless, using numerical computations, the authors identify many symmetric and asymmetric stationary solutions. It would be a worthwhile future question whether a classification via symmetries is available for the solutions of non-local chemotaxis models. The stability of the stationary states is studied via the free energy of the system, where coarsening of the systems stationary states corresponds to a step-wise decrease of the system's free energy.

In [16], a particle model for materials that are highly connected is proposed. Particles sufficiently close to each other may be linked via spring-like links. Link creation and destruction are random events. The model can be used to describe cellular tissues, mucins, polymers, or social networks. To study the evolution of systems comprised of a large number of particles, the authors derive a macroscopic description. Assuming that the network remodelling process is fast, this results in the limit being the aggregation equation. In [15], the authors study the stationary solutions by the way of bifurcation analysis. They find that the analytical and numerical results of the macroscopic model approximate well the dynamics of the microscopic model.

Part III
Non-local Equations with Boundary Conditions

Chapter 6
No-Flux Boundary Conditions for Non-local Operators

It is a challenge to define boundary conditions for non-local models on bounded domains. The periodic case, which we studied in the previous chapters, is an exception, since we can work with periodic extensions outside of the domain. However, no-flux conditions or Dirichlet or Robin boundary conditions need special attention.

In the literature, there are several ideas to deal with boundary conditions of non-local operators. For example, the inclusion of a potential that diverges close to the boundary can keep particles away from it [65, 191]. One could also simply cut off the part of the integral that reaches outside of the domain, as done, for example, in [100]. Another idea, often used in numerical simulations, is to introduce ghost points outside of the domain with certain symmetry properties [3].

In our case, we are interested in the motion of biological objects such as cells or bacteria; hence, we will formulate the non-local boundary conditions from a biological point of view. A cell can attach to the boundary, it could slip off it or be neutral relative to the boundary. Our boundary conditions need to be able to describe these cases.

Next we briefly summarize previous ideas in models for biological populations. Hillen et al. [100] considered the global existence of a non-local chemotaxis equation. To correctly define the non-local chemotaxis equation on a bounded domain, they limited the set over which the non-local operator (6.1) integrates

$$\overset{\circ}{\nabla}_R v(x) := \frac{n}{\omega_D(x)R} \int_{\mathbb{S}_D^{n-1}} \sigma v(x + R\sigma)\, d\sigma, \tag{6.1}$$

where $\mathbb{S}_D^{n-1}(x) = \{\sigma \in \mathbb{S}^{n-1} : x + \sigma R \in D\}$ and $\omega_D(x) = |\mathbb{S}_D^{n-1}(x)|$. The same approach is briefly discussed in [59]. While the non-local gradient as defined in Eq. (6.1) ensures that it is well-defined, it does not satisfy no-flux boundary conditions.

Xiang studied global bifurcations of the non-local chemotaxis equation using the global bifurcation analysis by Rabinowitz and Crandall (same approach as in Chap. 5) and modified the non-local gradient (6.1) in 1-D such that a no-flux condition is

A. Buttenschön, T. Hillen, *Non-Local Cell Adhesion Models*, CMS/CAIMS Books in Mathematics 1, https://doi.org/10.1007/978-3-030-67111-2_6

satisfied [192]. The construction assumed that two domains are in contact on the boundary; then using a reflection argument through $x = 0, L$, Xiang obtained

$$
\overset{\circ}{\nabla}_R v(x) := \frac{1}{2R} \begin{cases} v(x + R) - v(R - x) & \text{if } 0 \le x \le R, \\ v(x + R) - v(x - R) & \text{if } R < x < L - R \cdot \\ v(2L - x - R) - v(x - R) & \text{if } L - R \le x \le L \end{cases}
$$

A similar reflection approach is briefly discussed by Topaz et al. [177]. Another class of non-local models to study species aggregations are the so-called aggregation equations [65, 130, 191]. In these equations, the non-local term is the results of interactions between individuals. The interactions can be described using a potential energy. Recently, such equations have been studied on bounded domains [65, 191]. The resulting boundary conditions are very similar to ours.

6.1 Non-local No-Flux Boundary Conditions

In this chapter, we present a comprehensive theory for non-local boundary conditions of the adhesion model in one dimension. The key idea, as presented in [97], is to introduce a spatially dependent sampling domain called $E(x)$. It ensures that cells measure adhesive forces inside the domain and also on the domain boundary. Within this framework, we can define naive, no-flux, neutral, adhesive, and repellent boundary conditions, each relating to their own biological reality. We show results on local and global existence of solutions, steady states, and pattern formations.

We develop non-local boundary conditions for a one-dimensional domain $D = [0, L], L > 0$, where $L > 2R$ such that a single cell cannot touch both boundaries at the same time. Extensions into higher dimensions are not straightforward, and we will discuss them later. Let $u(x, t)$ be the density of a cell population at location $x \in D$ and time t, which diffuses and adheres to each other.

$$
u_t(x,t) = d u_{xx}(x,t) - \alpha \left(u(x,t) \int_{-R}^{R} h(u(x + r, t)) \Omega(r) \, dr \right)_x, \qquad (6.2)
$$

where d is the diffusion coefficient, R the cell sensing radius, α the strength of the homotypic adhesions, and $h(\cdot)$ a possibly nonlinear function describing the nature of the adhesive force. We denote the corresponding cell flux by

$$
J(x,t) := -d u_x(x,t) + \alpha u(x,t) \int_{-R}^{R} h(u(x + r, t)) \Omega(r) \, dr,
$$

which has two components: the diffusive flux and the adhesive flux. We use the cell flux to define boundary conditions. In a closed container, or a petri dish, it is reasonable to assume that no cells can leave or enter through the domain boundary;

hence, $J \cdot \mathbf{n} = 0$, where $\mathbf{n}$ denotes the outward pointing normal on ∂D. In our one-dimensional case, we have $\mathbf{n} = -1$ for $x = 0$ and $\mathbf{n} = +1$ for $x = L$. Hence, we require

$$J(0, t) = J(L, t) = 0. \tag{6.3}$$

As in [97], we consider two cases: (1) the diffusive flux and the adhesive flux are independent and (2) the diffusive flux and adhesive flux depend on each other.

6.1.1 Independent Fluxes

To satisfy the zero-flux condition (6.3), we assume that the diffusive and adhesive fluxes independently satisfy

$$u_x(0, t) = u_x(L, t) = 0, \qquad \mathcal{K}[u](0) = \mathcal{K}[u](L) = 0. \tag{6.4}$$

The condition for the diffusive flux is easily included in the mathematical problem formulation by restricting to the appropriate function space. For instance, given a function space X, we define a boundary operator $\mathcal{B}$ and construct a function space satisfying Neumann boundary conditions.

$$\mathcal{B}[u] = (u'(0), u'(L)), \qquad X_{\mathcal{B}} := X \cap N[\mathcal{B}].$$

The condition for the non-local term is more challenging, and in subsequent section, we show that this leads to a spatial dependence in the integration limits of the non-local term.

6.1.2 Dependent Flux

In certain situations, we can relax conditions (6.4) and allow non-zero adhesion fluxes on the boundary, i.e.,

$$\begin{aligned} du_x(0, t) &= u(0, t)\, \mathcal{K}[u](0, t), \\ du_x(L, t) &= u(L, t)\, \mathcal{K}[u](L, t). \end{aligned} \tag{6.5}$$

This means that the first derivative can be non-zero on the domain's boundary. This condition, however, is less amenable to mathematical analysis, as it shows the hyperbolic nature of the non-local drift term. We need to distinguish the incoming particle flux on the boundary from the outgoing part, i.e., at each location along the boundary, we would need to identify the direction of the net flux and stipulate boundary conditions only at influx points. It is not clear how this can be done. Hence, in this chapter, we focus on the independent flux case (Sect. 6.1.1).

In both cases, the definition of appropriate boundary conditions for the non-local operator $\mathcal{K}[u]$ is both a mathematical and modelling challenge. We need to ensure that the non-local operator $\mathcal{K}[u]$ is well-defined near the boundary and satisfies the no-flux boundary conditions. To determine the near boundary behavior, we look at the biology. When a cell encounters a boundary, it might attach to it, be repelled by it, or form a neutral attachment.

6.2 Naive Boundary Conditions

The simplest case to define sensible boundary conditions is a case used in [59, 100], where we remove any points of the sampling domain that fall outside the domain D.

$$E_0(x) = \{\, y \in V : x + y \in D \,\}.$$

We call this the naive case, as this does not employ any biological reasoning. We simply restrict the integration domain. The resulting operator is well-defined for all $x \in D$, but it does not necessarily satisfy the zero-flux condition (6.4). Indeed, this is easily observed when $\Omega(r) = \frac{1}{2R} \frac{r}{|r|}$ and when $\mathcal{K}$ is applied to a constant function $c > 0$. We observe that for $x \in [0, R)$, we have

$$\mathcal{K}[u] = \frac{1}{2R} \int_0^x -c \; dy + \frac{1}{2R} \int_x^{x+R} c \; dy = c \left(\frac{R - x}{2R} \right),$$

which does not go to zero as $x \to 0$. A similar computation can be made on the right boundary $x = L$. In this case, $\mathcal{K}$ satisfies the relaxed boundary conditions (6.5), and the adhesive flux is pointing into the domain. Thus, cells are repelled from the boundaries, and we would expect them to accumulate in the domain's interior. This is nicely demonstrated by the numerical solutions in this situation; see Fig. 6.1. We call boundary conditions with inward flow to be *repellent boundary conditions*.

6.3 No-Flux Boundary Conditions

In this section, we show how a more general sensing domain is constructed, based on biological principles, such that the corresponding non-local operator satisfies the independent boundary conditions (6.4).

In Sect. 2.1, we summarize the derivation of the non-local adhesion model from biological principles as developed in [31]. In the derivation, the non-local term resulted from the careful construction of the adhesive forces between cells. A key requisite for adhesive interaction between cells is the formation of adhesive bonds between cells. The magnitude of the adhesive force created at distance r from the cell center is a function of the density of adhesion bonds $N_b(r)$, available free space

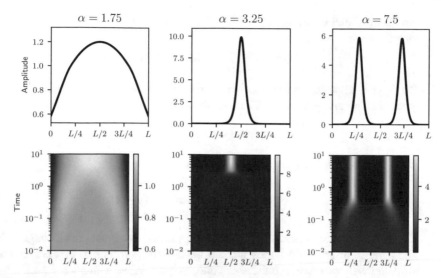

Fig. 6.1 Numerical solutions of Eq. (6.2), with the naive sensing domain $E_0(x)$. In the top row, we show the final solution profiles; below are the kymographs. (Left) $\alpha = 1.75$, (Middle) $\alpha = 3.25$, and (Right) $\alpha = 7.5$. The solutions feature several similarities to the periodic solutions in Fig. 5.11 (Sect. 5.5). For small α, we note the effect of the naive boundary conditions, repelling particles into the interior of the domain. For all shown α, there is no translational symmetry

$f(r)$, and the distance weight function $\Omega(r)$. In an unbounded domain, the explicit form of the non-local term is

$$\mathcal{K}[u](x) = \int_{-R}^{R} N_b(x + r) f(x + r) \Omega(r) \, dr.$$

Subsequently, we focus on the number of formed adhesion bonds N_b and assume that $f \equiv 1$. We make an additional assumption that the limits

$$\Omega^- := \lim_{x \to 0-} \Omega(x) \quad \text{and} \quad \Omega^+ := \lim_{x \to 0+} \Omega(x) \tag{6.6}$$

exist and are non-zero. Since Ω is an odd function, we have that

$$\Omega^- = -\Omega^+.$$

As discussed in the model derivation Sect. 2.1, the number of adhesion bonds $N_b(x+r)$ depends on the local cell population density $u(x+r)$. Hence, we write $N_b(x+r) = h(u(x + r))$ for an appropriate function $h \geq 0$. Often $h(u)$ is a linear function. For a bounded domain, the number of adhesion bonds is no longer homogeneous in space, since cells in the interior bind to neighboring cells, but cells close to the boundary bind to the boundary and to other cells. Hence, the function that describes the net adhesion bonds needs to explicitly depend on both, the location x of the probing cell and the sampling location $x + r$. Hence, we use $N_b = h(x, u(x + r))$ and obtain

$$\mathcal{K}[u](x) = \int_{E(x)} h(x, u(x + r))\Omega(r)\,dr. \tag{6.7}$$

To simplify the subsequent discussion, we introduce the following change of variables under the integral $y := x + r$. Then the sampling domain $E(x)$, in terms of y, is

$$\tilde{E}(x) = \{y \in D : |x - y| \leq R\}$$

With this, the non-local term (6.7) becomes

$$\mathcal{K}[u](x) = \int_{\tilde{E}(x)} h(x, u(y))\Omega(y - x)\,dy. \tag{6.8}$$

The boundary is solid. For this reason, cell protrusions cannot pass through it. Thus, the number of adhesion bonds formed beyond the wall has to be zero. Secondly, the solid wall may have repulsive or adhesive properties, which we describe by two additional functions h^0 and h^L, one for each boundary. As a result, the number of adhesion bonds for a cell located at $x \in D$ and binding at $y \in \tilde{E}(x)$ is given as

$$h(x, u(y)) = \begin{cases} H(u(y)) & \text{if } y \in \text{int}(D) \\ h^0(x)\delta(y) & \text{if } y = 0 \\ h^L(x)\delta(L - y) & \text{if } y = L \\ 0 & \text{else} \end{cases}.$$

Notice that the function $H(u(y))$ in the first case does not explicitly depend on x, since in the inner region we are in the homogeneous situation that cells interact with other cells within the sensing range. However, the functions $h^i(x)$ can be chosen such that the non-local term satisfies boundary conditions (6.4). First consider the non-local term $\mathcal{K}[u](x)$ defined in Eq. (6.8) in the interval $(0, R]$. There, the non-local term (6.8) can be decomposed as follows:

$$\mathcal{K}[u](x) = \int_x^{x+R} H(u(y))\Omega(y - x)\,dy + \int_0^x H(u(y))\Omega(y - x)\,dy + h^0(x)\Omega(-x).$$

Then, we choose the function $h^0(x)$ so that the boundary condition (6.4) is satisfied. We take the limit $x \to 0+$, using (6.6), and obtain

$$h^0(0) = \frac{-1}{\Omega^-} \int_0^R H(u(y))\Omega(y - x)\,dy = \frac{1}{\Omega^+} \int_0^R H(u(y))\Omega(y - x)\,dy. \tag{6.9}$$

Similarly, on the interval $[L - R, L)$, we have the following decomposition of the non-local term

$$\mathcal{K}[u](x) = \int_{x-R}^{x} H(u(y))\Omega(y - x)\,dy + \int_{x}^{L} H(u(y))\Omega(y - x)\,dy + h^{L}(x)\Omega(L - x).$$

To satisfy the no-flux boundary condition (6.4) at $x = L$, we consider the limit as $x \to L$ and find

$$h^{L}(L) = \frac{1}{\Omega^{-}} \int_{L-R}^{L} H(u(y))\Omega(y - x)\,dy. \tag{6.10}$$

It should be noted that if $\Omega^{-} = 0 = \Omega^{+}$, then we would need to require that the integrals in (6.9) and (6.10) are zero, which will not lead to a suitable boundary condition. Hence, the case of $\Omega^{-} = 0 = \Omega^{+}$ cannot be studied with the method presented here.

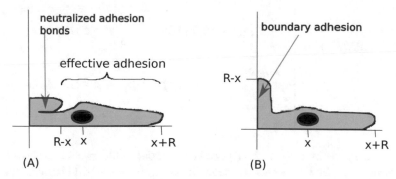

(A)   (B)

Fig. 6.2 (**a**) The filopodia of cell are reflected or stopped at the boundary. As a result, the cell starts to form adhesion bonds with itself, which are then not contributing to the net adhesion force. Note that only one cell is shown in this sketch. (**b**) The weak adhesive case. Cells make contact to the boundary in a well-balanced way, such that the net flux is still zero

The boundary condition (6.4) defines the value of the $h^{0}(x)$ at 0 and $h^{L}(x)$ at L. We still have a choice to define these functions in the interval of size R from the boundary. We will study several choices.

There are many possible choices of the function $h^{0}(x)$ such that condition (6.9) is satisfied. A natural choice is to assume that the adhesion near the boundary follows the same mechanism $H(u)$ as in the domain's interior. Therefore, it is natural to pick a function $h^{0}(x)$ that is continuous between $x = 0, R$. We choose

$$h^{0}(x) = \frac{1}{\Omega(x)} \int_{0}^{R-x} H(u(y))\Omega(y - x)\,dy, \; x \in [0, R], \tag{6.11}$$

where we also assume that $\Omega(r) \neq 0$ for all $r \in V\backslash\{0\}$, thus resulting in a non-local gradient for $x \in (0, R]$ of

$$\mathcal{K}[u](x) = \int_{R-x}^{x+R} H(u(y))\Omega(y - x)\,dy. \tag{6.12}$$

We make the same natural choice for the right boundary, i.e., we set $h^L(x)$ such that it satisfies condition (6.10) at $x = L$ and $h^L(L - R) = 0$. The resulting non-local operator for $x \in [L - R, L)$ is given by

$$\mathcal{K}[u](x) = \int_{x-R}^{2L-R-x} H(u(y))\Omega(y - x)\,\mathrm{d}y. \tag{6.13}$$

Combining the boundary non-local terms (6.12) and (6.13), and reverting the change of variables, we obtain

$$\mathcal{K}[u](x) = \begin{cases} \displaystyle\int_{R-2x}^{R} H(u(x + r))\Omega(r)\,\mathrm{d}r & \text{if } x \in (0, R] \\[2.5em] \displaystyle\int_{-R}^{R} H(u(x + r))\Omega(r)\,\mathrm{d}r & \text{if } x \in [R, L - R] \\[2.5em] \displaystyle\int_{-R}^{2L-R-2x} H(u(x + r))\Omega(r)\,\mathrm{d}r & \text{if } x \in [L - R, L) \\[2em] 0 & \text{if } x = 0, L \end{cases} \tag{6.14}$$

It is easy to observe that $\mathcal{K}$ is indeed continuous (which we will discuss in detail in Sect. 6.4). The $\Omega(r)$ terms preceding the integrals in Eq. (6.11) cancel out for $r \neq 0$.

This explicit form (6.14) allows us to write the integral operator as integral over an effective sensing domain. We define

$$f_1(x) = \begin{cases} R - 2x & \text{for } x \in [0, R) \\ -R & \text{for } x \in [R, L] \end{cases},$$

$$f_2(x) = \begin{cases} R & \text{for } x \in [0, L - R) \\ 2L - R - 2x & \text{for } x \in (L - R, L] \end{cases},$$

and the sampling domain

$$E_f(x) := [f_1(x), f_2(x)],$$

and write

$$\mathcal{K}[u] = \int_{E_f(x)} h(u(x + r))\Omega(r)\,\mathrm{d}r. \tag{6.15}$$

The sensing domain for this case is shown on the right of Fig. 6.5. Biologically, we interpret $E_f(x)$ as describing how the filopodia upon hitting the domain's boundary are reflected or stopped at the boundary. Consequently the cell starts to form adhesion bonds with itself, which are then not contributing to the net adhesion force (see also

Fig. 6.2). Numerical solutions of Eq. (6.2) for this particular choice of sensing domain are shown in Fig. 6.3.

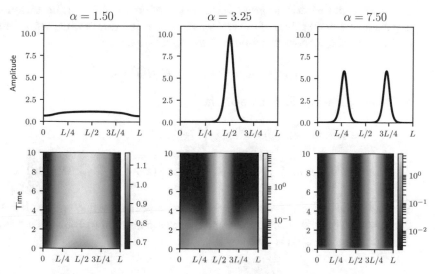

Fig. 6.3 Numerical simulations of Eq. (6.2), with the no-flux sensing domain $E_f(x)$. In the top row, we show the final solution profiles; below are the kymographs. (Left) $\alpha = 1.75$, (Middle) $\alpha = 3.25$, and (Right) $\alpha = 7.5$. The solutions feature several similarities to the periodic solutions in Fig. 5.11 (Sect. 5.5). For small α, we note the weakly repulsive effect of the boundaries. Finally, we note that for all values of α, the solutions have no translation symmetry

6.3.1 Approximate Steady States for No-Flux Non-local Term

In this section, we consider a singular perturbation of the steady state of the no-flux adhesion model for small adhesive strengths. The steady states of Eq. (6.2) with repellent non-local term $\mathcal{K}[u]$ are given by the solutions of the following equation.

$$u_{xx}(x,t) - \alpha \left(u(x,t) \int_{E_f(x)} u(x+r)\Omega(r)\,dr \right)_x = 0, \qquad (6.16a)$$

where $E_f(x)$ is no-flux sensing domain. As Eq. (6.2) exhibits mass conservation, we impose the following mass constraint on the solutions of Eq. (6.16a).

$$\mathcal{A}[u] = \bar{u}, \qquad (6.16b)$$

where $\mathbb{R} \ni \bar{u} > 0$ is the mass per unit length of population $u(x)$ in D. To be able to easily carry out the subsequent asymptotic expansion, we assume that the function $h(\cdot)$ under the integral in Eq. (6.16a) is linear (i.e., $h(u) = u$).

When $\alpha = 0$, the boundary conditions are given by the classical Neumann boundary conditions. It is then easy to see that Eq. (6.2) admits a constant steady-state solution $u \equiv \bar{u}$. When $\alpha \neq 0$, the situation is much more complicated.

Here we approximate the ground steady state of Eq. (6.16a) by using an asymptotic expansion for small values of α. Here we set $R = 1$. In the following, we assume that $\alpha = \epsilon$. Then we consider the following asymptotic expansion:

$$u(x) = u_0(x) + \epsilon u_1(x) + \epsilon^2 u_2(x) + O(\epsilon^3). \tag{6.17}$$

Substituting into Eq. (6.16a), we obtain

$$0 \approx \Big(u_0(x) + \epsilon u_1(x) + \epsilon^2 u_2(x)\Big)_{xx}$$
$$- \epsilon \left((u_0 + \epsilon u_1) \int_{-1}^{1} \Big(u_0 + \epsilon u_1 + \epsilon^2 u_2\Big)(x + r)\chi_{E_f(x)}(r)\Omega(r)\,dr\right)_x.$$

Separating the scales of ϵ, we obtain for the zeroth-order equation $(u_0)_{xx} = 0$. Due to the Neumann boundary conditions, and the mass constraint (6.16b), we have $u_0 \equiv \bar{u}$. The first-order equation is given by

$$(u_1)_{xx} - \left(u_0 \int_{-1}^{1} u_0(x + r)\chi_{E_f(x)}(r)\Omega(r)\,dr\right)_x = 0. \tag{6.18}$$

Equation (6.18) requires the solvability condition $\mathcal{A}[u_1] = 0$. Using the properties of the function $\Omega(r) = \frac{r}{|r|}\omega(r)$, we rewrite Eq. (6.18) as a piecewise condition

$$(u_1)_{xx} = \bar{u}^2 \begin{cases} 2\omega(1 - 2x) & \text{for } x \in [0, 1/2] \\ -2\omega(1 - 2x) & \text{for } x \in [1/2, 1] \\ 0 & \text{for } x \in [1, L - 1] \\ -2\omega(2L - 2x - 1) & \text{for } x \in [L - 1, L - 1/2] \\ 2\omega(2L - 2x - 1) & \text{for } x \in [L - 1/2, L] \end{cases}.$$

We solve this differential equation by integration, and after some algebra, we obtain:

$$
u_1(x) = \begin{cases}
u(0) + \bar{u}^2 \displaystyle\int_0^x \int_{1-2s}^1 \omega(r)\, dr\, ds & \text{for } x \in [0, 1/2] \\[2ex]
A - \bar{u}^2 \displaystyle\int_x^1 \int_{-1}^{1-2s} \omega(r)\, dr\, ds & \text{for } x \in [1/2, 1] \\[2ex]
A & \text{for } x \in [1, L-1] \\[2ex]
A - \bar{u}^2 \displaystyle\int_{L-1}^x \int_{2L-2s-1}^1 \omega(r)\, dr\, ds & \text{for } x \in [L-1, L-1/2] \\[2ex]
u(L) + \bar{u}^2 \displaystyle\int_x^L \int_{-1}^{2L-2s-1} \omega(r)\, dr\, ds & \text{for } x \in [L-1/2, L]
\end{cases}
$$

where

$$
A = \frac{\bar{u}^2}{L} \int_0^1 r\omega(r)\, dr
$$

and

$$
u(0) = u(L) = \bar{u}^2 \left[\frac{1-L}{L} \right] \int_0^1 r\omega(r)\, dr.
$$

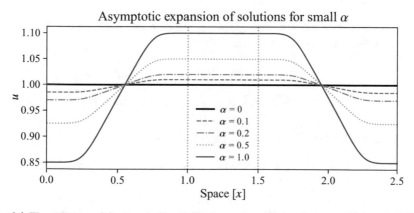

Fig. 6.4 The solution $u(x)$ given in Eq. (6.17) for various values of ϵ or α. The vertical black dotted lines denote the boundaries of the areas that are within one sensing radius of the domain boundary, at $x = R$ and $x = L - R$

For a visual depiction of the asymptotic expansion (6.17), see Fig. 6.4. The domain boundary in this case is weakly repellent. At steady state, cells experience adhesive forces from the domain interior and less from the domain boundary, hence leading to local minima at the boundary.

6.4 General Sampling Domain

We can generalize the above approach and define a general class of suitable sampling domains. These will include the naive case and the no-flux case and further cases for weak adhesion and repulsion as we discuss later. We define a general suitable domain of integration for the non-local term.

Definition 6.1 (Sensing Domain [97])

1. Two continuous functions $f_{1,2} : D \to \mathbb{R}$ define a suitable sampling domain

$$E(x) = [f_1(x), f_2(x)],$$

if they satisfy

  a. $-R \leq f_1(x) \leq f_2(x) \leq R$ for all $x \in [0, L]$.
  b. $f_1(x) = -R$ for $x \in [R, L]$
  c. $f_2(x) = R$ for $x \in [0, L - R]$.
  d. $f_1(x)$ and $f_2(x)$ are non-increasing and have uniformly bounded one-sided derivatives.

2. A suitable sampling domain $E(x)$ satisfies the second condition (6.4) if in addition

  e. $f_1(0) = R$ and $f_2(L) = -R$.

Two examples of suitable sampling domains are shown in Fig. 6.5. We view the non-local operator as function on $L^p(D)$.

$$\mathcal{K} : D \times L^p(D) \mapsto \mathbb{R}, \ p \geq 1.$$

$$\mathcal{K}[u(x)](x) := \int_{E(x)} h(u(x + r))\Omega(r)\,dr, \tag{6.19}$$

which satisfies the following assumptions:

  (A1) $\Omega(r) = \frac{r}{|r|}\omega(r)$, where $\omega(r) = \omega(-r)$,
  (A2) $\omega(r) \geq 0$, $\omega(R) = 0$,
  (A3) $\omega \in L^1(V) \cap L^\infty(V)$, and $\|\omega\|_{L^1([0,R])} = {}^1/_2$.
  (A4) The sampling domain $E(x)$ is suitable according to Definition 6.1.
  (A5) The adhesion function $h(u)$ is linearly bounded and differentiable and

$$|h(u)| \leq k_1(1 + |u|), \quad \text{and} \quad |h'(u)| \leq k_2, \quad k_1, k_2 > 0.$$

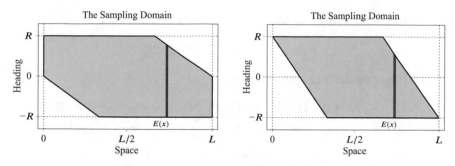

Fig. 6.5 Two examples of the spaces $D \times V$, with the spatial domain on the x-axis and the suitable sensing domain $E(x)$ on the y-axis. The shaded region is the set $\{(x, y) : x \in D, y \in E(x)\}$, and a sample sensing domain $E(x)$ of thickness dx is shown in the darker gray. Left: Here the sensing domain is of naive type. Right: Here the sensing domain is of no-flux type

6.4.1 Set Convergence

The spatial dependence in the integration limits can equivalently be viewed as introducing an indicator function under the integral. To establish the continuity and differentiability of $\mathcal{K}[u]$, we recall the calculus of indicator functions in general metric spaces. Let $(V, \mathcal{A}, \mu)$ be a sigma-finite measure space. Then let $A, B \in \mathcal{A}$, and we define the symmetric difference of A and B by

$$A \triangle B = (A \cup B) \setminus (A \cap B).$$

The function

$$d : \mathcal{A} \times \mathcal{A} \mapsto \mathbb{R}, \tag{6.20}$$

defined by

$$d(A, B) = \mu(A \triangle B) \quad \text{for } A, B \in \mathcal{A},$$

is a pseudometric and is known as the Fréchet-Nikodym metric [21], allowing us to view any sigma-finite measure space as pseudometric space. We can turn this into a metric by introducing the following equivalence relation:

$$X \sim Y \iff \mu(X \triangle Y) = 0. \tag{6.21}$$

It is easy to see that $A \sim B$ if and only if they differ by a set of measure zero. If we denote the set of equivalence classes of this relation by $\tilde{\mathcal{A}} = \mathcal{A}/\mu$, then the function (6.20) is extended by setting $d(\tilde{A}, \tilde{B}) = d(A, B)$ where $\tilde{A}, \tilde{B} \in \mathcal{A}/\mu$. This turns $\mathcal{A}/\mu$ into a complete metric space $\tilde{\mathcal{A}}$ [21, Theorem 1.12.6].

Lemma 6.1 *Let $(V, \mathcal{A}, \mu)$ be a measure space, with pseudometric (6.20) and $\tilde{\mathcal{A}}$ the corresponding metric space of equivalence classes of (6.21). Suppose that $E(x)$ is a suitable sampling domain as defined in Definition 6.1. Then $E(x)$ is continuous in $\tilde{\mathcal{A}}$ in the following sense: if $(x_n) \subset \mathbb{R}$ such that $x_n \to x$, then $\mu(E(x_n) \triangle E(x)) \to 0$.*

Proof We consider a sequence $x_n \to x \in D$ and define the sets

$$E_n := [f_1(x_n), f_2(x_n)]$$

The symmetric difference of $E(x)$ and E_n is

$$E_n \triangle E(x) = \{y \in V : \min(f_1(x), f_1(x_n)) \le y \le \max(f_1(x), f_1(x_n)),$$
$$\min(f_2(x), f_2(x_n)) \le y \le \max(f_2(x), f_2(x_n))\}$$

whose measure is computed to be

$$\mu(E_n \triangle E(x)) = |f_1(x_n) - f_1(x)| + |f_2(x_n) - f_2(x)|.$$

Since both $f_{1,2}$ are continuous, we have that

$$\mu(E_n \triangle E(x)) \to 0 \quad \text{as} \quad x_n \to x.$$

$\square$

Next we collect some properties of the indicator functions of symmetric differences. Suppose that $A, B \in \mathcal{A}$; then

$$\chi_{A \triangle B}(r) = \chi_A(r) + \chi_B(r) - 2\chi_{A \cap B}(r),$$
$$\chi_{A \cap B}(r) = \chi_A(r)\chi_B(r) = \min(\chi_A(r), \chi_B(r)),$$
$$\chi_{A \triangle B}(r) = |\chi_A(r) - \chi_B(r)|.$$

Having established the continuity of the sampling domains $E(x)$, we can consider the continuity of the integral operator (6.19).

6.4.2 Continuity near the Boundary

In this section, we explore continuity of the no-flux non-local operator defined in Eq. (6.19). To distinguish between the norms over the sensing domain $V = [-R, R]$ and the spatial domain D, we introduce norm notation that indicates the set over which the norm is taken. For example, in this subsection, the L^p norm over that set A is denoted as

$$|u|_{p,A} := \left(\int_A |u(x)|^p \, dx \right)^{1/p}.$$

Theorem 6.1 *Let $p \ge 1$, $u \in L^p(D)$, and assume that (A1)–(A5) are satisfied.*

1. The map $\left(r \mapsto \chi_{E(x)}(r)H(u(x+r))\Omega(r) \right) \in L^p(D)$, and there exists a constant $C > 0$ such that

$$|\chi_{E(x)}(r)H(u(x+r))|_{p,V} \le C(1 + |u|_{p,D}).$$

2. *The map* $\left(x \mapsto \mathcal{K}[u](x)\right) \in L^p(D)$, *and there exists a constant* $C > 0$ *such that*

$$|\mathcal{K}[u]|_{p,D} \leq C(1 + |u|_{p,D})$$

3. $\mathcal{K} : L^p(D) \mapsto L^p(D)$ *is continuous.*
4. *If* $u \in C^0(D)$, *then the map* $\left(x \mapsto \mathcal{K}[u](x)\right) \in C^0(D)$.

5. *The map* $\left(x \mapsto \mathcal{K}[u](x)\right) \in C^0(D)$.

Proof Fix $x \in D$. Then

$$\int_V |\chi_{E(x)}(r)H(u(x+r))\Omega(r)|^p \, dr \leq |\Omega|_\infty^p \int_D |H(u(x))|^p \, dx$$
$$\leq Ck_1^p \int_D (1 + |u(x)|)^p \, dx$$
$$\leq C(1 + |u|_{p,D})^p.$$

which proves item (1). To prove item (2), we use item (1) and estimate

$$|\mathcal{K}[u]|_{p,D}^p = \int_D \left| \int_V \chi_{E(x)(r)} H(u(x+r))\Omega(r) \, dr \right|^p \, dx$$
$$\leq \int_D \int_V \left(C(1 + |u|_{p,D})|\Omega|_{\infty,V} \right)^p \, dr \, dx$$
$$\leq \left(C(1 + |u|_{p,D})|\Omega|_{\infty,V} \right)^p |V||D|$$
$$\leq \left(C(1 + |u|_{p,D}) \right)^p.$$

To proof item (3), we consider a sequence $u_n \to u$ in $L^p(D)$ and use the Lipschitz continuity of H to estimate

$$|\mathcal{K}[u_n] - \mathcal{K}[u]|_{p,D} \leq \left| \int_D \chi_{E(x)}(r) \Big(H(u_n(x+r)) - H(u(x+r)) \Big) \Omega(r) \, dr \right|_{p,D}$$
$$\leq k_2 \left| \int_D \chi_{E(x)}(r)|u_n(x+r) - u(x+r)|\Omega(r) \, dr \right|_{p,D}$$
$$\leq C|u_n - u|_{p,D}$$

To show item (4), we consider $u \in C^0(D)$ and a sequence $(x_n) \subset D$ such that $x_n \to x \in D$. Then

$$\mathcal{K}[u](x) - \mathcal{K}[u](x_n) = \int_V \left\{ \chi_{E(x)}(r)H(u(x+r)) - \chi_{E(x_n)}H(u(x_n+r)) \right\} \Omega(r)\, dr$$

$$= \int_V \left\{ \chi_{E(x)}(r)H(u(x+r)) - \chi_{E(x_n)}H(u(x_n+r)) \right.$$

$$+ \chi_{E(x)\cap E(x_n)}(r)H(u(x_n+r)) - \chi_{E(x)\cap E(x_n)}(r)H(u(x_n+r))$$

$$+ \chi_{E(x)\cap E(x_n)}(r)H(u(x+r))$$

$$\left. - \chi_{E(x)\cap E(x_n)}(r)H(u(x+r)) \right\} \Omega(r)\, dr$$

$$= \int_V \left\{ H(u(x+r))\chi_{E(x)}(r)\left(\chi_{E(x)}(r) - \chi_{E(x_n)}(r)\right) \right.$$

$$+ H(u(x_n+r))\chi_{E(x_n)}(r)\left(\chi_{E(x_n)}(r) - \chi_{E(x)}(r)\right)$$

$$\left. + \chi_{E(x)\cap E(x_n)}(r)\left(H(u(x+r)) - H(u(x_n+r))\right) \right\} \Omega(r)\, dr.$$

Then, we obtain

$$|\mathcal{K}[u](x) - \mathcal{K}[u](x_n)| \le k_1(1+|u|_\infty)|\Omega|_\infty \left[2\int_V \left| \chi_{E(x)}(r) - \chi_{E(x_n)}(r) \right| dr \right.$$

$$\left. + \int_{E(x)\cap E(x_n)} |H(u(x+r)) - H(u(x_n+r))|\, dr \right].$$

Then, the integrals

$$\int_V |\chi_{E(x)}(r) - \chi_{E(x_n)}(r)|\, dr = \int_{E(x)\triangle E(x_n)} dr = \mu(E(x) \triangle E(x_n)).$$

By Lemma 6.1, we have that $\mu(E(x) \triangle E(x_n)) \to 0$ as $n \to \infty$. The second integral converges by the continuity of H and u.

Finally, to show item (5), we approximate $u \in L^p(D)$ with functions $u_n \in C^0(D)$, $u_n \to u$ in $L^p(D)$. Then

$$\mathcal{K}[u](x) - \mathcal{K}[u](x_n) = \mathcal{K}[u](x) - \mathcal{K}[u_m](x)$$

$$+ \mathcal{K}[u_m](x_n) - \mathcal{K}[u](x_n) + \mathcal{K}[u_m](x) - \mathcal{K}[u_m](x_n)$$

$$\le 2|u_m - u|_p |\Omega|_{\infty,V} + |\mathcal{K}[u_m](x) - \mathcal{K}[u_m](x_n)|.$$

The first term vanished due to the density of the smooth functions in $L^p(D)$ and item (3). The second terms vanished by item (4).   □

In summary, we have shown that whenever $u \in L^p(D)$ and the sensing domain $E(x)$ is continuous in the sense of Lemma 6.1, we have that the no-flux non-local term in Eq. (6.19) is continuous.

6.4.3 Differentiation near the Boundary

The integral limits of the non-local operator $\mathcal{K}$ are spatially dependent; hence, taking the spatial derivative of the non-local term (as done in the model) requires the differential of indicator functions, i.e., we consider a weak formulation.

Lemma 6.2 *In the sense of distributional derivatives, we find*

$$\mathcal{K}[u]'(x) = \frac{d}{dx}\left(\int_{E(x)} H(u(x+r))\Omega(r)\,dr\right)$$

$$= \int_{E(x)} \frac{d}{dx} H(u(x+r))\Omega(r)\,dr$$

$$+ f_2'(x)H(u(x+f_2(x)))\Omega(f_2(x)) - f_1'(x)H(u(x+f_1(x)))\Omega(f_1(x)). \quad (6.22)$$

Proof First, we use interval notation to rewrite the indicator function as

$$\chi_{E(x)}(r) = \chi_{(-R,f_2(x))}(r)\chi_{(f_1(x),R)}(r) = \mathcal{H}(r - f_1(x)\mathcal{H}(f_2(x) - r), \; r \in [-R, R].$$

We use the Heaviside function $\mathcal{H}$ and the Dirac-delta distribution δ to write

$$\frac{d}{dx}\chi_{E(x)}(r) = \frac{d}{dx}\{\mathcal{H}(r - f_1(x))\mathcal{H}(f_2(x) - r)\},$$

$$= \mathcal{H}(r - f_1(x))\delta(f_2(x) - r)f_2'(x) - \mathcal{H}(f_2(x) - r)\delta(r - f_1(x))f_1'(x).$$

Then

$$\int_V \frac{d}{dx}\chi_{E(x)}H(u(x+r))\Omega(r)\,dr$$

$$= \int_V \mathcal{H}(r - f_1(x))\delta(f_2(x) - r)f_2'(x)H(u(x+r))\Omega(r)\,dr$$

$$- \int_V \mathcal{H}(f_2(x) - r)\delta(r - f_1(x))f_1'(x)H(u(x+r))\Omega(r)\,dr$$

$$= \mathcal{H}(f_2(x) - f_1(x))\Big(f_2'(x)H(u(x+f_2(x)))\Omega(f_2(x)) - f_1'(x)H(u(x+f_1(x)))\Omega(f_1(x))\Big),$$

$$= f_2'(x)H(u(x+f_2(x)))\Omega(f_2(x)) - f_1'(x)H(u(x+f_1(x)))\Omega(f_1(x)).$$

Note that $f_2(x) \geq f_1(x)$ implies $\mathcal{H}(f_2(x) - f_1(x)) = 1$. Together with the product rule, we obtain (6.22). $\qquad \square$

Next we study the continuity properties of the derivative $\mathcal{K}[u]'$, and we use short-hand notations for the two parts of the derivative

$$\mathcal{K}[u]'(x) = D_1\,\mathcal{K}[u](x) + D_2\,\mathcal{K}[u](x)$$

$$D_1\,\mathcal{K}[u](x) = \int_{E(x)} \frac{d}{dx}H(u(x+r))\Omega(r)\,dr$$

$$D_2\,\mathcal{K}[u](x) = f_2'(x)H(u(x+f_2(x)))\Omega(f_2(x)) - f_1'(x)H(u(x+f_1(x)))\Omega(f_1(x)),$$

and study the two terms separately. The first term can be treated the same way as $\mathcal{K}[u]$ in Theorem 6.1.

Corollary 6.1 *Let* $p \geq 1$, $u \in W^{1,p}(D)$, *and assume that (A1)–(A5) are satisfied.*

1. *The map* $\left(r \mapsto \chi_{E(x)}(r)\frac{d}{dx}H(u(x+r))\Omega(r) \right) \in L^p(D)$, *and there exists a constant* $C > 0$ *such that*

$$\left| \chi_{E(x)}(r)\frac{d}{dx}H(u(x+r)) \right|_{p,V} \leq C(1 + |u|_{W^{1,p}}).$$

2. *The map* $\left(x \mapsto D_1 \mathcal{K}[u](x) \right) \in L^p(D)$, *and there exists a constant* $C > 0$ *such that*

$$|D_1 \mathcal{K}[u]|_{p,D} \leq C(1 + |u|_{W^{1,p}})$$

3. $D_1 \mathcal{K} : W^{1,p}(D) \to L^p(D)$ *is continuous.*
4. *The map* $\left(x \mapsto \mathcal{K}[u](x) \right) \in C^0(D)$.

Proof The proof follows directly from Theorem 6.1.   □

Let us consider $D_2 \mathcal{K}$.

Lemma 6.3 *Let* $p \geq 1$, $u \in W^{1,p}(D)$, *and assume that (A1)–(A5) are satisfied. The map*

$$x \mapsto D_2 \mathcal{K}[u](x)$$

is continuous for $x \in [0, L]$.

Proof Since the sampling slice $E(x) = [f_1(x), f_2(x)]$ is defined by suitable functions $f_1(x), f_2(x)$ given by Definition 6.1, the only possible points of discontinuity of $D_2 \mathcal{K}[u](x)$ are $x = R$ and $x = L - R$. Let us compute

$$\lim_{x \to R^-} D_2 \mathcal{K}[u](x) = - \lim_{x \to R^-} f_1'(x)H(u(u - R))\Omega(-R)$$
$$\lim_{x \to R^+} D_2 \mathcal{K}[u](x) = 0,$$

which is continuous, since we assumed f_1' is bounded and $\Omega(\pm R) = 0$. A limit for $x \to L - R$ leads to the condition $\Omega(R) = 0$.   □

Hence, the weak distributional derivative $\mathcal{K}[u]'(x)$ is in fact a classical derivative:

Corollary 6.2 *Let* $p \geq 1$, $u \in W^{1,p}(D)$, *and assume that (A1)–(A5) are satisfied; then* $\mathcal{K}[u]'(x)$ *is continuous in* $x \in [0, L]$. *Hence,* $\mathcal{K}[u]'(x)$ *is a classical derivative.*

6.5 Local and Global Existence

Having established the regularity of the non-local operator $\mathcal{K}[u]$ in Corollary 6.2, we are ready to state a result that ensures the global existence of solutions of Eq. (6.2).

Theorem 6.2 (Theorem from [97]) *Assume (A1)–(A5); then Eq. (6.2) has a unique solution in*

$$u \in C^0\left([0, \infty), H^2(0, L)\right).$$

In the case of independent fluxes, the proof of this theorem is given in [97]. The generalization of the existence result to the case of dependent fluxes is an open problem.

6.6 Neutral Boundary Conditions

A key feature of standard partial differential equations with homogeneous no-flux boundary conditions is the existence of constant steady states. For the non-local adhesion model (6.2), a constant steady state $\bar{u}$ must satisfy $\mathcal{K}[\bar{u}] = 0$. However, looking at the non-local operators that we defined above, naive case and weakly repellent case, then neither of these admits a constant steady state. To allow for constant steady states, we need to modify the non-local operator at the boundary to compensate for the repellent flux near the boundary.

We start with the non-local operator together with the no-flux sensing domain, and we add an additional correction function $c(x)$. We require that $c(0) = c(L) = 0$ such that $\mathcal{K}[u]$ still satisfies boundary condition (6.4).

$$\mathcal{K}[u](x) := \int_{E_f(x)} H(u(x + r))\Omega(r)\,dr - c(x). \tag{6.23}$$

We consider a constant $\bar{u} > 0$, and we want to choose $c(x)$ such that $\mathcal{K}[\bar{u}](x) = 0$ for any $x \in D$, where c is only non-zero near the boundary. Hence, we set

$$c(x) := \begin{cases} c^0(x) & \text{if } x \in [0, R) \\ 0 & \text{if } x \in [R, L - R] \, , \\ c^L(x) & \text{if } x \in (L - R, L] \end{cases} \tag{6.24}$$

To determine $c^0(x)$, we compute $\mathcal{K}[\bar{u}]$ in the boundary regions and solve for $c^0(x)$. We find that

$$c^0(x) := \begin{cases} H(\bar{u}) \displaystyle\int_{R-2x}^{R} \omega(r)\,dr & \text{if } x \in [0, R/2] \\ \\ H(\bar{u}) \displaystyle\int_{2x-R}^{R} \omega(r)\,dr & \text{if } x \in [R/2, R] \end{cases} .$$

To determine $c^L(x)$, we repeat this process on the right boundary and find that

$$
c^L(x) := \begin{cases}
-H(\bar{u}) \displaystyle\int_{2L-2x-R}^{R} \omega(r)\,dr & \text{if } x \in [L-R, L-R/2] \\[6mm]
-H(\bar{u}) \displaystyle\int_{-R}^{2L-2x-R} \omega(r)\,dr & \text{if } x \in [L-R/2, L].
\end{cases}
$$

With that definition of the function $c(x)$, we can rewrite the neutral repellent non-local operator $\mathcal{K}[u]$ as

$$
\mathcal{K}_n[u](x) := \int_{E_f(x)} [\, H(u(x+r)) - H(\bar{u}) \,]\, \Omega(r)\,dr. \tag{6.25}
$$

Since this operator respects constant solutions, we call it the *neutral version* of the repellent no-flux operator.

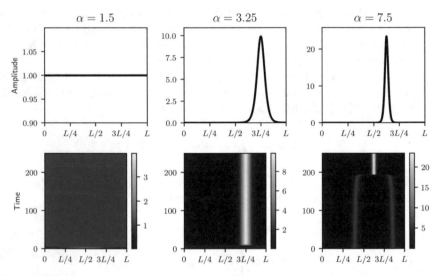

Fig. 6.6 Numerical solutions of Eq. (6.2), with the no-flux sensing domain $E_f(x)$, and a neutral non-local operator as constructed in Eq. (6.25). In the top row, we show the final solution profiles; below are the kymographs. (Left) $\alpha = 1.5$, (Middle) $\alpha = 3.25$, and (Right) $\alpha = 7.5$. Compared to the no-flux solutions in Fig. 6.3, the solutions here do not feel the boundary, and peaks may form anywhere in the domain

Note that this definition does not change the definition of the non-local term in the interior of the domain (i.e., when x is at least a distance of R from the boundary) since there $\mathcal{K}[\bar{u}] = 0$. It only changes the definition of the operator near the boundary. It is interesting to note that the neutral operator $\mathcal{K}_n$ compared to the original operator $\mathcal{K}$ can be written as

$$
\mathcal{K}_n[u] = \mathcal{K}[u] - \mathcal{K}[\bar{u}].
$$

This operation can be easily done for the other cases that we studied, and we get a neutral naive operator and a neutral weakly adhesive operator. In the periodic case, we have $\mathcal{K}[\bar{u}] = 0$; hence, the non-local operator in the periodic case is already neutral.

Some numerical solutions for three choices of α are shown in Fig. 6.6. Compared to the numerical solutions with no-flux boundary condition, the neutral boundary condition solutions do not feel the boundary (see Fig. 6.6).

6.7 Weakly Adhesive and Repulsive Boundary Conditions

The framework developed here can be used to explicitly model adhesion or repulsion by the domain boundary. For that we assume that the interaction force with the boundary is proportional to the extent of cell protrusions that attach to the boundary, which corresponds to the amount of cell protrusions that would reach out of the domain if there was no boundary (see Fig. 6.2b), for example, at $x \in (0, R)$. If the cell extends to $x - R$, then the interval $[x - R, 0)$ is outside of the domain. We assume that instead of leaving the domain, the protrusion interacts with the boundary, given boundary adhesion terms of the form

$$a^0(x) := \beta^0 \int_{-R}^{-x} \Omega(r) \, dr, \quad x \in [0, R)$$

$$a^L(x) := \beta^L \int_{L-x}^{R} \Omega(r) \, dr, \quad x \in (L - R, L]$$

where β^0 and β^L are constants of proportionality. $\beta^0, \beta^L > 0$ describes boundary adhesion, while $\beta^0, \beta^L < 0$ describes boundary repulsion.

In this case, we define the adhesion operator as linear combination of all relevant adhesive effects. Using indicator functions $\chi_A(r)$, we can write

$$\mathcal{K}[u](x, t) = \int_{E_0(x)} H(u(x + r, t))\Omega(r) \, dr$$

$$+ \beta^0 \chi_{[0,R]}(x) \int_{-R}^{-x} \Omega(r) \, dr + \beta^L \chi_{[L-R,L]}(x) \int_{L-x}^{R} \Omega(r) \, dr$$

$$= \int_{-R}^{R} \left(\chi_{E_0(x)} H(u(x + r, t)) + \beta^0 \chi_{[-R,-x]}(r) + \beta^L \chi_{[L-x,R]}(r) \right) \Omega(r) \, dr,$$

where we omitted the r dependence in the indicator functions for brevity. Here $E(x)$ is any suitable sampling domain as defined in Definition 6.1. Further we note that whenever

$$\beta^0 = \frac{1}{2} \int_{E(0)} H(u(r, t))\Omega(r) \, dr,$$

a similar expression can be found for β^L; then $\mathcal{K}[u]$ satisfies condition (6.4).

For this choice of a non-local operator, we explore the solutions of Eq. (6.2) numerically. Numerical solutions for different values of α are shown in Fig. 6.7 for $\beta^{0,L} > 0$ and Fig. 6.8 for $\beta^{0,L} < 0$.

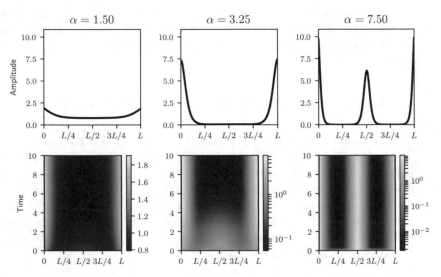

Fig. 6.7 Numerical solutions of Eq. (6.2), with the naive sensing domain $E_0(x)$. The non-local operator is modified as given in Eq. (6.26), here $\beta^0, \beta^L = 2$. In the top row, we show the final solution profiles; below are the kymographs. (Left) $\alpha = 1.5$, (Middle) $\alpha = 3.25$, and (Right) $\alpha = 7.5$. In these simulations, the weakly adhesive nature of the boundaries attracts cells and leads to aggregations on the boundary

6.8 Conclusion

In this chapter, we considered the non-local cell adhesion model (6.2) on a bounded domain with no-flux boundary conditions. The presence of the boundary meant that the definition of the non-local operator in the boundary region (within one sensing radius of the boundary) had to be revisited. There are many different ways to construct non-local operators that satisfy the no-flux boundary condition (6.3), and we developed a general framework through the sampling domain $E(x)$.

We recall while discussing the periodic case in Sect. 2.1, we argued that the non-local term can be seen as cell polarization. This understanding also helps to visualize the effect of the boundary conditions discussed here. The cell polarization points in the direction in which the most new adhesion bonds were formed. In the no-flux case, more adhesion bonds are formed in the domain inside than along the boundary; hence, cells at the boundary are facing inward. This is consistent with biological observations of cells approaching solid boundaries [149].

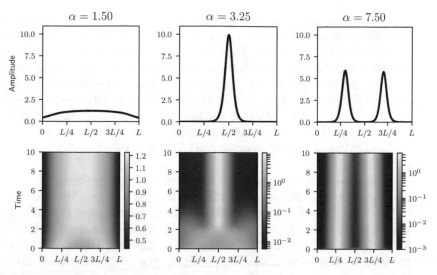

Fig. 6.8 Numerical solutions of Eq. (6.2), with the naive sensing domain $E_0(x)$. The non-local operator is modified as given in Eq. (6.26), here $\beta^0, \beta^L = -1$. In the top row, we show the final solution profiles; below are the kymographs. (Left) $\alpha = 1.5$, (Middle) $\alpha = 3.25$, and (Right) $\alpha = 7.5$. In these simulations, the weakly repulsive nature of the boundaries pushes cells away, leading to cell-free boundaries

Case	$\mathcal{K}[u]$	$f_1(x)$	$f_2(x)$
Periodic	$\int_V h(u)\Omega\,dr$	$-R$	R
Naive	$\int_{E_0(x)} h(u)\Omega\,dr$	$\begin{cases} -x \ I_1 \\ -R \ I_2 \end{cases}$	$\begin{cases} R, \quad I_3 \\ L-x, \ I_4 \end{cases}$
Non-flux	$\int_{E_f(x)} H(u)\Omega\,dr$	$\begin{cases} R-2x, \ I_1 \\ -R, \quad I_2 \end{cases}$	$\begin{cases} R, \quad\quad\quad I_3 \\ 2L-R-2x, \ I_4 \end{cases}$
Weakly adhesive	$\int_{E(x)} H(u)\Omega\,dr + a_0 + a_L$ $a_0 = -2h^0(0)\int_{-R}^{-x}\Omega(r)\,dr$ $a_L = 2h^L(L)\int_{L-x}^{R}\Omega(r)\,dr$	f_1 =naive	f_2 = naive
Neutral	$\mathcal{K}_n[u] = \mathcal{K}[u] - \mathcal{K}[\bar{u}]$	Any of the above cases	

Table 6.1 The different cases of suitable boundary conditions on $[0, L]$. The sensing domain is defined as $E(x) = [f_1(x), f_2(x)]$. The abbreviations I_1, I_2, I_3, and I_4 stand for $x \in [0, R]$, $x \in (R, L]$, $x \in [0, L-R]$, and $x \in (L-R, L]$, respectively

In Table 6.1, we summarize the specific boundary conditions that we studied here. The periodic case is fully translational symmetric. The naive case and the no-flux case are weakly repellent. The (neutral) non-local operator switches between a repulsive and attractive behavior, depending on whether the population size is above or below the reference mass $\bar{u}$. The neutral non-local operator has the important property that $\mathcal{K}[\bar{u}] = 0$ for all $x \in D$. For this reason, the bifurcation approach used in Chap. 5 might be applied to the neutral case. It is interesting to note that only recently Watanabe et al. also used the comparison with the mass per unit length to obtain solutions of Burger's equation on a bounded domain with no-flux boundary conditions [185].

In the case of the repellent non-local operator, we asymptotically approximated the steady states for small values of adhesion strength α. A more detailed exploration of the steady states of these equations is hindered by the fact that there is no constant steady state that exists for all values of α. This means that the approach pioneered by Rabinowitz [156] (used in Chap. 5) cannot be directly applied. In addition, even the introduction of the neutral non-local operator does not improve the situation. This is because due to the spatial dependence of the non-local operator, we are unable to embed the no-flux solutions into a periodic problem (see, for instance, [72] where this is done for reaction-diffusion systems).

In the construction of the non-local operators that include the no-flux boundary conditions, we had freedom to choose their behavior in the boundary region (within one sensing radius of the boundary). From biological experiments, it is known that the cell polarization adapts when cells encounter physical boundaries [149]. Responsible for this adaptation are the intracellular signalling networks. In [31], we argued that the non-local operator is a model of cell polarization. Thus, a natural extension would be to explicitly include the intracellular chemical network in a detailed multi-scale model of cell adhesion.

Mathematically novel is the non-local operator in which the integration limits are spatially dependent, whose mathematical properties we investigated further. The spatial dependence of the integration limits posed a particular challenge, since properties such as continuity require a notion of convergence of sets (integration domain). For this reason, we make use of the Fréchet-Nikodym metric (see Sect. 6.4.1). Using this metric, we extended the estimates of Sect. 3.2, to the non-local operator with spatially dependent limits of integration. Differentiation of this non-local operator was equally made more challenging by the spatially dependent integration limits. Using the theory of distributions, we computed the non-local term's weak derivative, which coincides with the classical derivative if the integration kernel $\Omega(\cdot)$ is zero on the boundary of the sensing domain (∂V).

A heat equation with non-local Robin-type boundary conditions was studied by Arendt et al. in [9]. In their case, the boundary condition is linear and can be seen as a perturbation of the heat equation semigroup. In our case, even for linear $h(u)$, the non-local term $\alpha(u\mathcal{K}[u])_x$ is nonlinear, and Arendt's perturbation approach would not work.

Another challenge is the higher-dimensional case. Boundary conditions need to distinguish between transversal and tangential effects, and combinations of slip, adhesion, repulsion, and friction are biologically possible. We hope that the current formalism can form a good basis to investigate higher-dimensional adhesion models on bounded domains.

Chapter 7
Discussion and Future Directions

The central building block to include adhesive interactions between cells in reaction-advection-diffusion models of tissues is to use a non-local term. The simplest scalar non-local partial differential equation including such a term is given by

$$u_t(x,t) = u_{xx}(x,t) - \alpha \left(u(x,t) \int_{-R}^{R} h(u(x+r,t))\Omega(r)\,dr \right), \qquad (7.1)$$

where $u(x,t)$ denotes the density of a cell population which adheres to itself, $h(u)$ measures the adhesive strength with the background population, and $\Omega(r)$ describes the distribution of adhesive sites on the cell membrane and on cell protrusions. This served as our prototypical example to develop the mathematical tools to investigate the steady-state structure of such an equation and their stability. In addition, having biological applications, we posed the question how to correctly formulate no-flux boundary conditions for such non-local models and how these boundary conditions affect the equation's steady states.

In Chap. 3, we study the properties of the non-local operator $\mathcal{K}$. We observe that the non-local operator is a generalization of the classical first-order derivative and shares many of its properties. In fact, as the sensing radius tends to zero, the non-local operator converges to the classical derivative. It also has eigenvalues that are strongly reminiscent of the classical derivative. Unsurprisingly, however, since it is a non-local operator, it is indeed compact, and its eigenvalues accumulate at zero.

The steady states of the periodic problem are discussed in Part II. The advantage of the periodic problem is that it allows us to discuss the steady states in absence of any boundary effects. Since the trivial solution exists for all values of α, we use the abstract global bifurcation results pioneered by Rabinowitz [156], to identify bifurcation points of non-trivial solutions. Since the non-local operator inherits the same symmetry properties as the classical first-order derivative, we can classify the globally existing solution branches, using the symmetries of the adhesion equation. The symmetries of each bifurcation branch allow us to show that each branch is contained in a separate space of spiky functions (i.e., piecewise monotone functions having a fixed number of zeros). Due to the symmetries, the locations of the peaks and valleys of these solutions are fixed and uniformly spaced. Due to the symmetries, all bifurcations are of pitchfork type, and their criticality is solely determined by the properties of the kernel $\omega(r)$ in the non-local term.

The limitation of our main global bifurcation result, namely, the restriction to linear functions $h(u) = u$ and constant kernel $\omega(r)$, is invitation for much future work. Indeed, it is not trivial to improve on either limitation. The main challenge one encounters trying to improve upon both limitations is generalizing the "non-local" maximum principle. The current proof relies on the relationship between the area function of a solution $u(x)$ and the non-local operator $\mathcal{K}$.

The next step in the analysis of the dynamics of this scalar non-local equation is to study its time-dependent solutions. This is of both mathematical and biological interest. Questions from a modelling perspective include: Can cell aggregates merge? Do there exist meta-stable solutions? What happens to aggregates as $t \to \infty$. In mathematical terms, we are seeking a full characterization of the attractor of the scalar non-local adhesion equation. We believe that some of the mathematical methods we developed in the analysis here will be invaluable in answering such questions in the future.

7.1 Further Thoughts

A different kind of result has been shown for semilinear scalar parabolic equation for which it is known that the so-called lap number (i.e., the number of zeros of function) is a non-increasing function with respect to time [125]. This means that solutions can only become simpler as time progresses and all complexity must be present in the initial conditions. In [126], this result was extended to scalar equations on the unit circle S^1. There however we have two possibilities: either the solutions approach a set of fixed points or the solutions resemble a rotating wave. Further, the rotating wave solution becomes impossible when the equation's nonlinearity possesses a reflection symmetry in its derivative dependence. In the case of the scalar non-local equation (7.1), extensive numerical exploration suggest that the number of spikes (i.e., number of zeros of the solution's derivatives) is non-increasing as a function of time.

The Chafee-Infante equation is one of the most studied nonlinear scalar reaction-diffusion equations. Non-local extensions to the Chafee-Infante problem have been studied in [159] and similar non-local problems in [49, 68–70]. The non-local term interacts with the regular spectrum of the Laplacian by moving some eigenvalues [68, 69]. Depending on the severity of this "scrambling," this leads to the stabilization or destabilization of spatial patterns and to changes in flow direction on the global attractor [159]. These spectral results are similar to the result we observed for our canonical example of a non-local equation (7.1). Whether similar results hold for the global attractor remains to be seen.

For scalar nonlinear reaction diffusion equations, it is proven in [88, 96] that if ϕ and ψ are two hyperbolic equilibria, then $W^u(\phi)$ is transversal to $W^s(\psi)$. This means that dynamical changes on the global attractor due to variations in the non-linearity can only occur due to a bifurcation. Does a similar result apply to our canonical non-local equation (7.1)?

Using the adhesion potential in Sect. 2.6, we made a strong connection to the existing literature on the aggregation equations and the Mckean-Vlasov model [36, 37]. In fact for $h(u) = u$, our model is a Mckean-Vlasov model. This connection makes variational methods available for the analysis of stability and qualitative behavior. We indicated some of these connections throughout the text, and a full detailed analysis of this relationship seems a promising way for future research.

7.2 Systems and Higher Dimensions

The scalar non-local adhesion equation studied in this monograph served as a kind of canonical example, on which we developed the mathematical tools necessary to tackle more complicated and more biologically realistic models. Originally the development of the non-local adhesion model was motivated by the well-studied phenomena of cell sorting, in which cells of two distinct populations can form different tissue structures depending solely on their adhesive properties. These effects are crucial during the development of organisms. The simplest model of cell sorting considered in [10] is

$$
\begin{aligned}
u_t &= u_{xx} - \alpha_{uu} \left(u(x,t) \, \mathcal{K}[u](x,t) \right)_x - \alpha_{uv} \left(u(x,t) \, \mathcal{K}[v](x,t) \right) \\
v_t &= v_{xx} - \alpha_{vv} \left(v(x,t) \, \mathcal{K}[v](x,t) \right)_x - \alpha_{vu} \left(v(x,t) \, \mathcal{K}[u](x,t) \right),
\end{aligned}
\tag{7.2}
$$

where $u(x,t)$ and $v(x,t)$ denote the densities of the two cell populations and α_{ij} denotes the strength of the adhesive strength of cell population i to j. Similarly, to the case of scalar reaction-diffusion equations, we expect the set of possible steady states to be much more complicated in this case (see, for instance, [73, 140], in which both symmetries were crucial to understand the equation's steady states and time-dependent solutions). Numerical simulations in these cases hint that symmetries remain important. As discussed in the Introduction, systems of the form (7.2) have been used extensively in cancer modelling.

Extensions to higher spatial dimensions would enable us to explore situations that are more closely connected with experimental situations, such as models of wound healing and tissue formation. Some results can be taken from the bifurcation analysis for the n-dimensional Mckean-Vlasov model [37].

Yet another different avenue would be to rigorously study the dynamics of this equation with spatio-temporally varying adhesion coefficients. Such variations due to either environmental changes or genetic or epigenetic variations are highly important in biological settings. For instance, it is clear that upon wounding cells change their behavior to close the wound and then return to their initial state. Another situation in which this is relevant is cancer growth and cancer metastasis, and this has been demonstrated by numerical studies in [56]. Other extensions include considering volume-filling effects and nonlinear diffusion, which both have been shown to be important in biological applications [39, 133].

In conclusion, in this monograph, we made significant progress in understanding the mathematical properties of non-local models of cell adhesion. We proposed a framework to derive such models from an underlying stochastic random walk, studied the steady states spawning through bifurcations from the constant solution, and finally considered the non-local cell adhesion model on a bounded domain with no-flux boundary conditions. The analysis of these non-local models is just beginning, and there are many worthwhile avenues of further research: most notably identification of secondary bifurcations, analysis of the time-dependent solutions of Eq. (1.1), several cell populations, higher space dimensions, and inclusion of intracellular signalling networks.

References

1. M. Adioui, O. Arino, N. El Saadi, A nonlocal model of phytoplankton aggregation. Nonlinear Anal. Real World Appl. **6**(4), 593–607 (2005)
2. B. Alberts, *Molecular Biology of the Cell: Reference Edition*, 5th edn. (Garland Science, New York, 2008)
3. M.B. Allen, E.L. Isaacson, *Numerical Analysis for Applied Science* (Wiley, New York, 1997)
4. W. Alt, Models for mutual attraction and aggregation of motile individuals, in *Mathematics in Biology and Medicine*. Lecture Notes in Biomathematics (Springer, Berlin, 1985), pp. 33–38
5. W. Alt, Degenerate diffusion equations with drift functionals modelling aggregation. Nonlinear Anal. Theory Methods Appl. **9**(8), 811–836 (1985)
6. L. Ambrosio, N. Gigli, G. Savaré, *Gradient Flows in Metric Spaces and in the Space of Probability Measures* (Birkhäuser, Basel, 2005)
7. V. Andasari, M.A.J. Chaplain, Intracellular modelling of cell-matrix adhesion during cancer cell invasion. Math. Model. Nat. Phenom. **7**(1), 29–48 (2012)
8. A.R.A. Anderson, A hybrid mathematical model of solid tumour invasion: the importance of cell adhesion. Math. Med. Biol. **22**(2), 163–86 (2005)
9. W. Arendt, S. Kunkel, M. Kunze, Diffusion with nonlocal Robin boundary conditions. J. Math. Soc. Jpn. **70**(4), 1523–1556 (2018)
10. N.J. Armstrong, K.J. Painter, J.A. Sherratt, A continuum approach to modelling cell-cell adhesion. J. Theor. Biol. **243**(1), 98–113 (2006)
11. N.J. Armstrong, K.J. Painter, J.A. Sherratt, Adding adhesion to a chemical signaling model for somite formation. Bull. Math. Biol. **71**(1), 1–24 (2009)
12. P.B. Armstrong, Light and electron microscope studies of cell sorting in combinations of chick embryo neural retina and retinal pigment epithelium. Wilhelm Roux Arch. Entwickl. Org. **168**(2), 125–141 (1971)
13. R. Bailo, J.A. Carrillo, J. Hu, Fully discrete positivity-preserving and energy-dissipating schemes for aggregation-diffusion equations with a gradient-flow structure. Commun. Math. Sci. **18**(5), 1259–1303 (2020)
14. R. Bailo, J.A. Carrillo, H. Murakawa, M. Schmidtchen, Convergence of a fully discrete and energy-dissipating finite-volume scheme for aggregation-diffusion equations (2020). Preprint, arXiv:2002.10821
15. J. Barré, J.A. Carrillo, P. Degond, D. Peurichard, E. Zatorska, Particle interactions mediated by dynamical networks: assessment of macroscopic descriptions. J. Nonlinear Sci. **28**(1), 235–268 (2018)
16. J. Barré, P. Degond, E. Zatorska, Kinetic theory of particle interactions mediated by dynamical networks. Multiscale Model. Simul. **15**(3), 1294–1323 (2017)
17. G. Bell, Models for the specific adhesion of cells to cells. Science **200**(4342), 618–627 (1978)

18. A.J. Bernoff, C.M. Topaz, Nonlocal aggregation models: a primer of swarm equilibria. SIAM Rev. **55**(4), 709–747 (2013)
19. D.A. Beysens, G. Forgacs, J.A. Glazier, Cell sorting is analogous to phase ordering in fluids. Proc. Natl. Acad. Sci. **97**(17), 9467–9471 (2000)
20. V. Bitsouni, M.A.J. Chaplain, R. Eftimie, Mathematical modelling of cancer invasion: The multiple roles of tgf/β pathway on tumour proliferation and cell adhesion. Math. Models Methods Appl. Sci. **27**(10), 1929–1962 (2017)
21. V.I. Bogachev, *Measure Theory* (Springer, Berlin, 2007)
22. I. Borsi, A. Fasano, M. Primicerio, T. Hillen, A non-local model for cancer stem cells and the tumour growth paradox. Math. Med. Biol. **34**(1), 59–75 (2017)
23. B. Brandolini, P. Freitas, C. Nitsch, C. Trombetti, Sharp estimates and saturation phenomena for a nonlocal eigenvalue problem. Adv. Math. **228**(4), 2352–2365 (2011)
24. N.F. Britton, *Reaction-Diffusion Equations and Their Applications to Biology* (Academic Press, London, 1986)
25. G.W. Brodland, The differential interfacial tension hypothesis (DITH): a comprehensive theory for the self-rearrangement of embryonic cells and tissues. J. Biomech. Eng. **124**(2), 188 (2002)
26. G.W. Brodland, H.H. Chen, The mechanics of cell sorting and envelopment. J. Biomech. **33**(7), 845–51 (2000)
27. B. Buffoni, J. Toland, *Analytic Theory of Global Bifurcation: An Introduction*, vol. 55 (Princeton University Press, Princeton, 2003)
28. P.L. Buono, R. Eftimie, Symmetries and pattern formation in hyperbolic versus parabolic models of self-organised aggregation. J. Math. Biol. **71**(4), 847–881 (2015)
29. M. Burger, M. Di Francesco, S. Fagioli, A. Stevens, Sorting phenomena in a mathematical model for two mutually attracting/repelling species. SIAM J. Math. Anal. **50**(3), 3210–3250 (2018)
30. A. Buttenschoen, T. Kolokolnikov, M.J. Ward, J. Wei, Cops-on-the-dots: the linear stability of crime hotspots for a 1-d reaction-diffusion model of urban crime. Eur. J. Appl. Math. **31**(5), 1–47 (2019)
31. A. Buttenschön, T. Hillen, A. Gerisch, K.J. Painter, A space-jump derivation for non-local models of cell-cell adhesion and non-local chemotaxis. J. Math. Biol. **76**(1), 429–456 (2018)
32. H.M. Byrne, M.A.J. Chaplain, Modelling the role of cell-cell adhesion in the growth and development of carcinomas. Math. Comput. Model. **24**(12), 1–17 (1996)
33. H.M. Byrne, D. Drasdo, Individual-based and continuum models of growing cell populations: a comparison. J. Math. Biol. **58**, 657–687 (2009)
34. J. Calvo, J. Campos, V. Caselles, O. Sánchez, J. Soler, Flux-saturated porous media equations and applications. EMS Surv. Math. Sci. **2**(1), 131–218 (2015)
35. J.A. Carrillo, X. Chen, Q. Wang, Z.A. Wang, L. Zhang, Phase transitions and bump solutions of the Keller–Segel model with volume exclusion. SIAM J. Appl. Math. **80**(1), 232–261 (2020)
36. J.A. Carrillo, A. Chertock, Y. Huang, A finite-volume method for nonlinear nonlocal equations with a gradient flow structure (2014). Preprint, arXiv:1402.4252
37. J.A. Carrillo, R.S. Gvalani, G.A. Pavliotis, A. Schlichting, Long-time behaviour and phase transitions for the Mckean–Vlasov equation on the torus. Arch. Ration. Mech. Anal. **235**(1), 635–690 (2020)
38. J.A. Carrillo, Y. Huang, M. Schmidtchen, Zoology of a nonlocal cross-diffusion model for two species. SIAM J. Appl. Math. **78**(2), 1078–1104 (2018)
39. J.A. Carrillo, H. Murakawa, M. Sato, H. Togashi, O. Trush, A population dynamics model of cell-cell adhesion incorporating population pressure and density saturation (2019). Preprint, arXiv:1901.02919
40. M.A.J. Chaplain, M. Lachowicz, Z. Szymanska, D. Wrzosek, Mathematical modelling of cancer invasion: the importance of cell-cell adhesion and cell-matrix adhesion. Math. Models Methods Appl. Sci. **21**(04), 719–743 (2011)

41. H.H. Chen, G.W. Brodland, Cell-level finite element studies of viscous cells in planar aggregates. J. Biomech. Eng. **122**(4), 394–401 (2000)
42. R.H. Chisholm, T. Lorenzi, A. Lorz, A. Larsen, L. Almeida, A. Escargueil, J. Clairambault, Emergence of drug tolerance in cancer cell populations: an evolutionary outcome of selection, non-genetic instability and stress-induced adaptation. Cancer Res. **75**, 930–939 (2015)
43. A. Coddington, N. Levinson, *Theory of Ordinary Differential Equations*. International Series in Pure and Applied Mathematics (McGraw-Hill, New York, 1955)
44. R. Courant, Ein allgemeiner Satz zur Theorie der Eigenfunktione selbstadjungierter Differentialausdrücke. Nach. Ges. Wiss. Göttingen Math.-Phys. Kl. 81–84 (1923)
45. M.G. Crandall, P.H. Rabinowitz, Nonlinear Sturm-Liouville eigenvalue problems and topological degree. J. Math. Mech. **19**(12), 1083–1102 (1970)
46. M.G. Crandall, P.H. Rabinowitz, Bifurcation from simple eigenvalues. J. Funct. Anal. **8**, 321–340 (1971)
47. M.G. Crandall, P.H. Rabinowitz, Bifurcation, perturbation of simple eigenvalues, and linearized stability. Arch. Ration. Mech. Anal. **52**(2), 161–180 (1973)
48. E.N. Dancer, Bifurcation from simple eigenvalues and eigenvalues of geometric multiplicity one. Bull. Lond. Math. Soc. **34**(5), 533–538 (2002)
49. F.A. Davidson, N. Dodds, Spectral properties of non-local differential operators. Appl. Anal. **85**(6–7), 717–734 (2006)
50. F.A. Davidson, N. Dodds, Spectral properties of non-local uniformly-elliptic operators. Electron. J. Differ. Equ. **126**, 1–15 (2006)
51. F.A. Davidson, N. Dodds, Existence of positive solutions due to non-local interactions in a class of nonlinear boundary value problems. Methods Appl. Anal. **14**(1), 15–28 (2007)
52. M. Delgado, I.B.M. Duarte, A. Suarez, Nonlocal problem arising from the birth-jump processes. Proc. R. Soc. Edinb. Sect. A Math. **149**(2), 1–23 (2018)
53. P. Deuflhard, *Newton Methods for Nonlinear Problems: Affine Invariance and Adaptive Algorithms*, vol. 35 (Springer Science & Business Media, Berlin, 2011)
54. E. Doedel, H.B. Keller, J.P. Kernevez, Numerical analysis and control of bifurcation problems (i): bifurcation in finite dimensions. Int. J. Bifurcat. Chaos **1**(03), 493–520 (1991)
55. E. Doedel, H.B. Keller, J.P. Kernevez, Numerical analysis and control of bifurcation problems (ii): bifurcation in infinite dimensions. Int. J. Bifurcat. Chaos **1**(04), 745–772 (1991)
56. P. Domschke, D. Trucu, A. Gerisch, M.A.J. Chaplain, Mathematical modelling of cancer invasion: implications of cell adhesion variability for tumour infiltrative growth patterns. J. Theor. Biol. **361C**, 41–60 (2014)
57. D. Drasdo, S. Höhme, A single-cell-based model of tumor growth in vitro: monolayers and spheroids. Phys. Biol. **2**(3), 133–147 (2005)
58. A. Ducrot, P. Magal, Asymptotic behavior of a non-local diffusive logistic equation. SIAM J. Math. Anal. **46**, 1731–1753 (2014)
59. J. Dyson, S.A. Gourley, R. Villella-Bressan, G.F. Webb, Existence and asymptotic properties of solutions of a nonlocal evolution equation modeling cell-cell adhesion. SIAM J. Math. Anal. **42**(4), 1784–1804 (2010)
60. L. Edelstein-Keshet, *Mathematical Models in Biology* (SIAM, Philadelphia, 2005)
61. R. Eftimie, G. de Vries, M.A. Lewis, Complex spatial group patterns result from different animal communication mechanisms. Proc. Natl. Acad. Sci. USA **104**(17), 6974–6979 (2007)
62. R. Eftimie, G. De Vries, M.A. Lewis, F. Lutscher, Modeling group formation and activity patterns in self-organizing collectives of individuals. Bull. Math. Biol. **69**(5), 1537–1565 (2007)
63. M. Fabian, P. Habala, P. Hájek, V.M. Santalucía, J. Pelant, V. Zizler, *Functional Analysis and Infinite-Dimensional Geometry* (Springer, Berlin, 2001)
64. P.E. Farrell, A. Birkisson, S.W. Funke, Deflation techniques for finding distinct solutions of nonlinear partial differential equations. SIAM J. Sci. Comput. **37**(4), A2026–A2045 (2015)
65. R.C. Fetecau, M. Kovacic, Swarm equilibria in domains with boundaries. SIAM J. Appl. Dyn. Syst. **16**(3), 1260–1308 (2017)

66. P.C. Fife, *Mathematical Aspects of Reacting and Diffusing Systems*, vol. 28 (Springer, Berlin, 1979)
67. R.A. Foty, M.S. Steinberg, Cadherin-mediated cell-cell adhesion and tissue segregation in relation to malignancy. Int. J. Dev. Biol. **48**(5–6), 397–409 (2004)
68. P. Freitas, Bifurcation and stability of stationary solutions of nonlocal scalar reaction-diffusion equations. J. Dyn. Differ. Equ. **6**(4), 613–629 (1994)
69. P. Freitas, A nonlocal Sturm–Liouville eigenvalue problem. Proc. R. Soc. Edinb. Math. **124**(01), 169–188 (1994)
70. P. Freitas, M. Vishnevskii, Stability of stationary solutions of nonlocal reaction-diffusion equations in m-dimensional space. Differ. Integral Equ. **13**(1–3), 265–288 (2000)
71. P. Friedl, R. Mayor, Tuning collective cell migration by cell-cell junction regulation. Cold Spring Harb. Perspect. Biol. **9**, a029199 (2017)
72. H. Fujii, M. Mimura, Y. Nishiura, A picture of the global bifurcation diagram in ecological interacting and diffusing systems. Physica D **5**(1), 1–42 (1982)
73. H. Fujii, Y. Nishiura, Global bifurcation diagram in nonlinear diffusion systems, in *Northholland Mathematics Studies* (Elsevier, Amsterdam, 1983), pp. 17–35
74. A. Gerisch, Numerical methods for the simulation of taxis diffusion reaction systems. Ph.D. Thesis, Martin-Luther-Universitat Halle-Wittenberg, 2001
75. A. Gerisch, On the approximation and efficient evaluation of integral terms in PDE models of cell adhesion. IMA J. Numer. Anal. **30**(1), 173–194 (2010)
76. A. Gerisch, M.A.J. Chaplain, Mathematical modelling of cancer cell invasion of tissue: local and non-local models and the effect of adhesion. J. Theor. Biol. **250**(4), 684–704 (2008)
77. A. Gerisch, K.J. Painter, Mathematical modeling of cell adhesion and its applications to developmental biology and cancer invasion, in *Cell Mechanics: From Single Scale-Based Models to Multiscale Modelling*, ed. by A. Chauviére, L. Preziosi, C. Verdier (CRC Press, Boca Raton, 2010), pp. 319–350
78. D. Gilbarg, N.S. Trudinger, *Elliptic Partial Differential Equations of Second Order*. Classics in Mathematics (Springer, Berlin, 1983)
79. S.F. Gilbert, M.J.F. Barresi, *Developmental Biology*, 12th edn. (Oxford University Press, Oxford, 2020)
80. J.A. Glazier, F. Graner, Simulation of the differential adhesion driven rearrangement of biological cells. Phys. Rev. E **47**(3), 2128–2154 (1993)
81. I. Gohberg, S. Goldberg, M.A. Kaashoek, *Classes of Linear Operators Vol. I*, vol. 49 (Birkhäuser, Basel, 1990)
82. M. Golubitsky, I. Stewart, *The Symmetry Perspective*. Progress in Mathematics (Birkhäuser, Basel, 2002)
83. F. Graner, Can surface adhesion drive cell-rearrangement? Part I: Biological cell-sorting. J. Theor. Biol. **164**(4), 455–476 (1993)
84. F. Graner, J.A. Glazier, Simulation of biological cell sorting using a two-dimensional extended Potts model. Phys. Rev. Lett. **69**(13), 2013–2016 (1992)
85. F. Graner, Y. Sawada, Can surface adhesion drive cell rearrangement? Part II: a geometrical model. J. Theor. Biol. **164**(4), 477–506 (1993)
86. M.L. Graves, J.A. Cipollone, P. Austin, E.M. Bell, J.S. Nielsen, C.B. Gilks, K.M. McNagny, C.D. Roskelley, The cell surface mucin podocalyxin regulates collective breast tumor budding. Breast Cancer Res. **18**(1), 11 (2016)
87. J.M. Halbleib, W.J. Nelson, Cadherins in development: Cell adhesion, sorting, and tissue morphogenesis. Genes Dev. **20**(23), 3199–3214 (2006)
88. J.K. Hale, *Asymptotic Behavior of Dissipative Systems* (American Mathematical Soc., Providence, 1988)
89. D. Hanahan, R.A. Weinberg, The hallmarks of cancer. Cell **100**(1), 57–70 (2000)
90. D. Hanahan, R.A. Weinberg, Hallmarks of cancer: the next generation. Cell **144**(5), 646–674 (2011)
91. A.K. Harris, Is cell sorting caused by differences in the work of intercellular adhesion? A critique of the Steinberg hypothesis. J. Theor. Biol. **61**(2), 267–285 (1976)

92. C.R. Harris, K.J. Millman, S.J. van der Walt, R.A. Gommers, P.I. Virtanen, D. Cournapeau, E. Wieser, J. Taylor, S. Berg, N.J. Smith, Array programming with numpy. Nature **585**(7825), 357–362 (2020)

93. T.J. Healey, Global bifurcations and continuation in the presence of symmetry with an application to solid mechanics. SIAM J. Math. Anal. **19**(4), 824–840 (1988)

94. T.J. Healey, H.J. Kielhöfer, Symmetry and nodal properties in the global bifurcation analysis of quasi-linear elliptic equations. Arch. Ration. Mech. Anal. **113**(4), 299–311 (1991)

95. T.J. Healey, H.J. Kielhöfer, Preservation of nodal structure on global bifurcating solution branches of elliptic equations with symmetry. J. Differ. Equ. **106**(1), 70–89 (1993)

96. D.B. Henry, Some infinite-dimensional morse-smale systems defined by parabolic partial differential equations. J. Differ. Equ. **59**(2), 165–205 (1985)

97. T. Hillen, A. Buttenschön, Nonlocal adhesion models for microorganisms on bounded domains. SIAM J. Appl. Math. **80**(1), 382–401 (2020)

98. T. Hillen, H. Enderling, P. Hahnfeldt, The tumor growth paradox and immune system-mediated selection for cancer stem cells. Bull. Math. Biol. **75**(1), 161–184 (2013)

99. T. Hillen, B. Greese, J. Martin, G. de Vries, Birth-jump processes and application to forest fire spotting. J. Biol. Dyn. **9**(Suppl. 1), 104–127 (2015)

100. T. Hillen, C. Painter, K.J. Schmeiser, Global existence for chemotaxis with finite sampling radius. Discrete Contin. Dyn. Syst. Ser. B **7**(1), 125–144 (2006)

101. T. Hillen, K.J. Painter, A user's guide to pde models for chemotaxis. J. Math. Biol. **58**(1–2), 183–217 (2009)

102. T. Hillen, M. Painter, K.J. Winkler, Global solvability and explicit bounds for a non-local adhesion model. Eur. J. Appl. Math. **29**, 645–684 (2018)

103. T. Hillen, A. Potapov, The one-dimensional chemotaxis model: global existence and asymptotic profile. Math. Methods Appl. Sci. **27**(15), 1783–1801 (2004)

104. S. Hoehme, D. Drasdo, Biomechanical and nutrient controls in the growth of mammalian cell populations. Math. Popul. Stud. **17**(3), 166–187 (2010)

105. D. Horstmann, From 1970 until present : the Keller-Segel model in chemotaxis and its consequences. Jahresber. Dtsch. Math. Vereinigung **105**(3), 103–165 (2003)

106. S.B. Hsu, J. López-Gómez, L. Mei, M. Molina-Meyer, A nonlocal problem from conservation biology. SIAM J. Math. Anal. **46**(6), 4035–4059 (2014)

107. B.D. Hughes, *Random Walks and Random Environments: Random Walks* (Oxford Science Publications, Clarendon Press, Oxford, 1995)

108. T. Ikeda, Standing pulse-like solutions of a spatially aggregating population model. Jpn. J. Appl. Math. **2**(1), 111–149 (1985)

109. T. Ikeda, T. Nagai, Stability of localized stationary solutions. Jpn. J. Appl. Math. **4**(1), 73–97 (1987)

110. D. Iron, M.J. Ward, A metastable spike solution for a nonlocal reaction-diffusion model. SIAM J. Appl. Math. **60**(3), 778–802 (2000)

111. K.L. Johnson, K. Kendall, A.D. Roberts, Surface energy and the contact of elastic solids. Proc. R. Soc. A Math. Phys. Eng. Sci. **324**(1558), 301–313 (1971)

112. K. Kang, T. Kolokolnikov, M.J. Ward, The stability and dynamics of a spike in the one-dimensional Keller-Segel model. IMA J. Appl. Math. **72**(2), 140–162 (2005)

113. R. Klages, *Anomalous Transport: Foundations and Applications* (Wiley, Hoboken, 2008)

114. H. Knútsdóttir, E. Pálsson, L. Edelstein-Keshet, Mathematical model of macrophage-facilitated breast cancer cells invasion. J. Theor. Biol. **357**, 184–199 (2014)

115. T. Kolokolnikov, J. Wei, Stability of spiky solutions in a competition model with cross-diffusion. SIAM J. Appl. Math. **71**(4), 1428–1457 (2011)

116. M. Kot, M.A. Lewis, M.G. Neubert, Integrodifference equations, in *Encyclopedia of Theoretical Ecology*, ed. by A. Hastings, L. Gross (University of California Press, Berkeley, 2012), pp. 382–384

117. J. López-Gómez, *Linear Second Order Elliptic Operators* (World Scientific Publishing Company, Singapore, 2013)

118. J.S. Lowengrub, H.B. Frieboes, F. Jin, Y.-L. Chuang, X. Li, P. Macklin, S.M. Wise, V. Cristini, Nonlinear modelling of cancer: bridging the gap between cells and tumours. Nonlinearity **23**(1), R1–R91 (2010)

119. J. López-Gómez, On the structure and stability of the set of solutions of a nonlocal problem modeling ohmic heating. J. Dyn. Differ. Equ. **10**(4), 537–566 (1998)

120. J. López-Gómez, *Spectral Theory and Nonlinear Functional Analysis*. Chapman & Hall/CRC Research Notes in Mathematics Series (Taylor & Francis, Hoboken, 2001)

121. J. López-Gómez, Global bifurcation for Fredholm operators. Rend. Ist. Mat. Univ. Trieste Int. J. Math. **48**, 539–564 (2016)

122. J. López-Gómez, C. Mora-Corral, *Algebraic Multiplicity of Eigenvalues of Linear Operators*. Operator Theory: Advances and Applications, vol. 177 (Birkhäuser, Basel, 2007)

123. P. Macklin, S. McDougall, A.R.A. Anderson, M.A.J. Chaplain, V. Cristini, J.S. Lowengrub, Multiscale modelling and nonlinear simulation of vascular tumour growth. J. Math. Biol. **58**(4–5), 765–798 (2009)

124. J. Martin, T. Hillen, The spotting distribution of wildfires. Appl. Sci. **6**(6), 177 (2016)

125. H. Matano, Nonincrease of the lap-number of a solution for a one-dimensional semilinear parabolic equation. J. Fac. Sci. Univ. Tokyo 1A **29**, 401–441 (1982)

126. H. Matano, Asymptotic behavior of solutions of semilinear heat equations on S^1, in *Nonlinear Diffusion Equations and Their Equilibrium States II* (Springer, Cham, 1988), pp. 139–162

127. P. McMillen, S.A. Holley, Integration of cell-cell and cell-ECM adhesion in vertebrate morphogenesis. Curr. Opin. Cell Biol. **36**, 48–53 (2015)

128. A. Meurer, C.P. Smith, M. Paprocki, O. Čertík, S.B. Kirpichev, M. Rocklin, A. Kumar, S. Ivanov, J.K. Moore, S. Singh, Sympy: symbolic computing in python. PeerJ. Comput. Sci. **3**, e103 (2017)

129. A. Mogilner, Modelling spatio-angular patterns in cell biology. Ph.D. Thesis, University of British Columbia, 1995

130. A. Mogilner, L. Edelstein-Keshet, A non-local model for a swarm. J. Math. Biol. **38**(6), 534–570 (1999)

131. J.C.M. Mombach, J.A. Glazier, R.C. Raphael, M. Zajac, Quantitative comparison between differential adhesion models and cell sorting in the presence and absence of fluctuations. Phys. Rev. Lett. **75**(11), 2244–2247 (1995)

132. A. Muntean, F. Toschi, *Collective Dynamics from Bacteria to Crowds* (Springer, Basel, 2004)

133. H. Murakawa, H. Togashi, Continuous models for cell-cell adhesion. J. Theor. Biol. **374**, 1–12 (2015)

134. J.D. Murray, *Mathematical Biology* (Springer, Berlin, 1993)

135. T. Nagai, Some nonlinear degenerate diffusion equations with a nonlocally convective term in ecology. Hiroshima Math. J. **13**(1), 165–202 (1983)

136. T. Nagai, M. Mimura, Asymptotic behavior for a nonlinear degenerate diffusion equation in population dynamics. SIAM J. Appl. Math. **43**(3), 449–464 (1983)

137. T. Nagai, M. Mimura, Some nonlinear degenerate diffusion equations related to population dynamics. J. Math. Soc. Jpn. **35**(3), 539–562 (1983)

138. M.G. Neubert, H. Caswell, Demography and dispersal: calculation and sensitivity analysis of invasion speed for structured populations. Ecology **81**(6), 1613–1628 (2000)

139. M.A. Nieto, The ins and outs of the epithelial to mesenchymal transition in health and disease. Annu. Rev. Cell Dev. Biol. **27**, 347–376 (2011)

140. Y. Nishiura, Global structure of bifurcating solutions of some reaction-diffusion systems. SIAM J. Math. Anal. **13**(4), 555–593 (1982)

141. A. Okubo, S.A. Levin, *Diffusion and Ecological Problems: Modern Perspectives*, vol. 14 (Springer Science & Business Media, New York, 2001)

142. H.G. Othmer, S.R. Dunbar, W. Alt, Models of dispersal in biological systems. J. Math. Biol. **26**(3), 263–298 (1988)

143. H.G. Othmer, T. Hillen, The diffusion limit of transport equations II: Chemotaxis equations. SIAM J. Appl. Math. **62**, 1222–1250 (2002)

144. C. Ou, Y. Zhang, Traveling wavefronts of nonlocal reaction-diffusion models for adhesion in cell aggregation and cancer invasion. Can. Appl. Math. Q. **21**(1), 21–62 (2013)
145. J.A. Owen, J. Punt, S.A. Stranford, *Kuby Immunology* (W. H. Freeman, New York, 2013)
146. J.M. Painter, K.J., Bloomfield, J.A. Sherratt, A. Gerisch, A nonlocal model for contact attraction and repulsion in heterogeneous cell populations. Bull. Math. Biol. **77**(6), 1132–1165 (2015)
147. K.J. Painter, T. Hillen, Spatio-temporal chaos in a chemotaxis model. Physica D **240**(4–5), 363–375 (2011)
148. K.J. Painter, N.J. Armstrong, J.A. Sherratt, The impact of adhesion on cellular invasion processes in cancer and development. J. Theor. Biol. **264**(3), 1057–1067 (2010)
149. A. Paksa, J. Bandemer, B. Hoeckendorf, N. Razin, K. Tarbashevich, S. Minina, D. Meyen, A. Biundo, S.A. Leidel, N. Peyriéras, N.S. Gov, P.J. Keller, E. Raz, Repulsive cues combined with physical barriers and cell-cell adhesion determine progenitor cell positioning during organogenesis. Nat. Commun. **7**, 1–14 (2016)
150. E. Palsson, H.G. Othmer, A model for individual and collective cell movement in dictyostelium discoideum. Proc. Natl. Acad. Sci. USA **97**(19), 10448–10453 (2000)
151. A.J. Perumpanani, J.A. Sherratt, J. Norbury, H.M. Byrne, Biological inferences from a mathematical model for malignant invasion. Invasion Metastasis **16**(4–5), 209–221 (1996)
152. A.B. Potapov, T. Hillen, Metastability in chemotaxis models. J. Dyn. Differ. Equ. **17**(2), 293–330 (2005)
153. M.H. Protter, H.F. Weinberger, *Maximum Principles in Differential Equations* (Springer, Berlin, 1984)
154. A. Puliafito, L. Hufnagel, P. Neveu, S. Streichan, A. Sigal, D.K. Fygenson, B.I. Shraiman, Collective and single cell behavior in epithelial contact inhibition. Proc. Natl. Acad. Sci. USA **109**(3), 739–744 (2012)
155. P.H. Rabinowitz, Nonlinear Sturm-Liouville problems for second order ordinary differential equations. Commun. Pure Appl. Math. **23**(1), 970, 939–961 (1970)
156. P.H. Rabinowitz, Some global results for nonlinear eigenvalue problems. J. Funct. Anal. **513**, 487–513 (1971)
157. I. Ramis-Conde, M.A.J. Chaplain, A.R.A. Anderson, Mathematical modelling of cancer cell invasion of tissue. Math. Comput. Model. **47**(5–6), 533–545 (2008)
158. I. Ramis-Conde, D. Drasdo, A.R.A. Anderson, M.A.J. Chaplain, Modeling the influence of the e-cadherin-beta-catenin pathway in cancer cell invasion: a multiscale approach. Biophys. J. **95**(1), 155–165 (2008)
159. J.B. Raquepas, J.D. Dockery, Dynamics of a reaction–diffusion equation with nonlocal inhibition. Physica D **134**(1), 94–110 (1999)
160. J.C. Robinson, *Infinite-Dimensional Dynamical Systems: An Introduction to Dissipative Parabolic PDEs and the Theory of Global Attractors*, vol. 28 (Cambridge University Press, Cambridge, 2001)
161. R. Schaaf, Stationary solutions of chemotaxis systems. Trans. Am. Math. Soc. **292**(2), 531–531 (1985)
162. M. Schienbein, K. Franke, H. Gruler, Random walk and directed movement: comparison between inert particles and self-organized molecular machines. Phys. Rev. E **49**(6), 5462–5471 (1994)
163. D.K. Schlüter, I. Ramis-Conde, M.A.J. Chaplain, Computational modeling of single-cell migration: the leading role of extracellular matrix fibers. Biophys. J. **103**(6), 1141–1151 (2012)
164. D.K. Schlüter, I. Ramis-Conde, M.A.J. Chaplain, Multi-scale modelling of the dynamics of cell colonies: insights into cell-adhesion forces and cancer invasion from in silico simulations. J. R. Soc. Interface **12**(103), 20141080 (2015)
165. M. Scianna, L. Preziosi, *Cellular Potts Models: Multiscale Extensions and Biological Applications*. Chapman & Hall/CRC Mathematical and Computational Biology (Taylor & Francis, Boca Raton, 2013)

166. J.A. Sherratt, S.A. Gourley, N.J. Armstrong, K.J. Painter, Boundedness of solutions of a non-local reaction-diffusion model for adhesion in cell aggregation and cancer invasion. Eur. J. Appl. Math. **20**(01), 123–144 (2009)

167. J. Shi, X. Wang, On global bifurcation for quasilinear elliptic systems on bounded domains. J. Differ. Equ. **246**(7), 2788–2812 (2009)

168. B.D. Sleeman, M.J. Ward, J.C. Wei, The existence and stability of spike patterns in a chemotaxis model. SIAM J. Appl. Math. **65**(3), 790–817 (2005)

169. J. Smoller, *Shock Waves and Reaction—Diffusion Equations*, vol. 258 (Springer, Berlin, 1994)

170. M.S. Steinberg, Reconstruction of tissues by dissociated cells. Science **141**(3579), 401–408 (1963)

171. M.S. Steinberg, Does differential adhesion govern self-assembly processes in histogenesis? Equilibrium configurations and the emergence of a hierarchy among populations of embryonic cells. J. Exp. Zool. **173**(4), 395–433 (1970)

172. M.S. Steinberg, Differential adhesion in morphogenesis: a modern view. Curr. Opin. Genet. Dev. **17**(4), 281–286 (2007)

173. A. Stevens, H.G. Othmer, Aggregation, blowup, and collapse: the ABC's of taxis in reinforced random walks. SIAM J. Appl. Math. **57**(4), 1044–1081 (1997)

174. J. Wei, T. Kolokolnikov, M.J. Ward, The stability of steady-state hot-spot patterns for a reaction-diffusion model of urban crime. Discrete Continuous Dyn. Syst. B **19**, 1373 (2014)

175. H.B. Taylor, A. Khuong, Z. Wu, Q. Xu, R. Morley, L. Gregory, A. Poliakov, W.R. Taylor, D.G. Wilkinson, Cell segregation and border sharpening by eph receptor–ephrin-mediated heterotypic repulsion. J. R. Soc. Interface **14**(132), 20170338 (2017)

176. J.P. Taylor-King, R. Klages, R.A. Van Gorder, Fractional diffusion equation for an n-dimensional correlated Levy walk. Phys. Rev. E **94**(1), 012104 (2016)

177. C.M. Topaz, A.L. Bertozzi, M.A. Lewis, A nonlocal continuum model for biological aggregation. Bull. Math. Biol. **68**(7), 1601–1623 (2006)

178. L.N. Trefethen, *Approximation Theory and Approximation Practice*, vol. 164 (SIAM, Philadelphia, 2019)

179. W.H. Tse, M.J. Ward, Hotspot formation and dynamics for a continuum model of urban crime. Eur. J. Appl. Math. **27**(3), 583–624 (2016)

180. S. Turner, J.A. Sherratt, Intercellular adhesion and cancer invasion: a discrete simulation using the extended Potts model. J. Theor. Biol. **216**(1), 85–100 (2002)

181. S. Turner, J.A. Sherratt, K.J. Painter, N. Savill, From a discrete to a continuous model of biological cell movement. Phys. Rev. E **69**(2), 021910 (2004)

182. N.G. Van Kampen, *Stochastic Processes in Physics and Chemistry*, 3rd edn. (North-Holland Personal Library, Elsevier Science, Amsterdam, 2007)

183. P. Virtanen, R.A. Gommers, T.E. Oliphant, M. Haberland, T. Reddy, D. Cour napeau, E. Burovski, P. Peterson, W. Weckesser, J. Bright, SciPy 1.0: fundamental algorithms for scientific computing in python. Nat. Methods **17**(3), 261–272 (2020)

184. X. Wang, Q. Xu, Spiky and transition layer steady states of chemotaxis systems via global bifurcation and helly's compactness theorem. J. Math. Biol. **66**, 1241–1266 (2013)

185. S. Watanabe, S. Matsumoto, T. Higurashi, N. Ono, Burgers equation with no-flux boundary conditions and its application for complete fluid separation. Physica D **331**, 1–12 (2016)

186. J. Wei, Existence and stability of spikes for the Gierer-Meinhardt system. *Handbook of Differential Equations: Stationary Partial Differential Equations*, ed. by M. Chipot, vol. 5 (North-Holland, Amsterdam, 2008), pp. 487–585

187. R.A. Weinberg, *The Biology of Cancer* (Garland Science, New York, 2013)

188. R. Weiner, B.A. Schmitt, H. Podhaisky, Rowmap—a row-code with Krylov techniques for large stiff odes. Appl. Numer. Math. **25**, 303–319 (1997)

189. H.V. Wilson, On some phenomena of coalescence and regeneration in sponges. J. Elisha Mitchell Sci. Soc. **23**(4), 161–174 (1907)

190. G.B. Wright, M. Javed, H. Montanelli, L.N. Trefethen, Extension of chebfun to periodic functions. SIAM J. Sci. Comput. **37**(5), C554–C573 (2015)

191. L. Wu, D. Slepčev, Nonlocal interaction equations in environments with heterogeneities and boundaries. Commun. Partial Differ. Equ. **40**(7), 1241–1281 (2015)
192. T. Xiang, A study on the positive nonconstant steady states of nonlocal chemotaxis systems. Discrete Contin. Dyn. Syst. Ser. B **18**(9), 2457–2485 (2013)
193. X. Zhang, L. Mei, On a nonlocal reaction-diffusion-advection system modeling phyto-plankton growth with light and nutrients. Discrete Contin. Dyn. Syst. Ser. B **17**(1), 221–243 (2011)
194. A. Zygmund, *Trigonometric Series*. Cambridge Mathematical Library, 3rd edn. (Cambridge University Press, Cambridge, 2002)

Index

Printed in the United States
by Baker & Taylor Publisher Services